MW01103445

IEEE Guide for Maintenance, Operation, and Safety of Industrial and Commercial Power Systems

Sponsor

Power Systems Engineering Committee
of the
Industrial and Commercial Power Systems Department
of the
IEEE Industry Applications Society

Approved 30 October 1998

IEEE-SA Standards Board

Abstract: Guidelines for the numerous personnel who are responsible for safely operating and maintaining industrial and commercial electric power facilities are provided. This guide provides plant engineers with a reference source for the fundamentals of safe and reliable maintenance and operation of industrial and commercial electric power distribution systems.
Keywords: electrical hazards, electrical maintenance, electrical safety program, fire protection, grounding, infrared, inspection, maintenance, operation protective devices, record keeping, safety single-line diagram, testing

Grateful acknowledgment is made to the following organization for having granted permission to reprint material in this document as listed below:

Electro • test inc. (eti), 725 Powell Avenue, SW Suite A, Renton, WA 98055-1212, USA for Figures 6-1 through 6-5.

First Printing
December 1998
SH94676

The Institute of Electrical and Electronics Engineers, Inc.
345 East 47th Street, New York, NY 10017-2394, USA

Print: ISBN 0-7381-1423-5 SH94676
PDF: ISBN 0-7381-1424-3 SS94676

IEEE Standards documents are developed within the IEEE Societies and the Standards Coordinating Committees of the IEEE Standards Association (IEEE-SA) Standards Board. Members of the committees serve voluntarily and without compensation. They are not necessarily members of the Institute. The standards developed within IEEE represent a consensus of the broad expertise on the subject within the Institute as well as those activities outside of IEEE that have expressed an interest in participating in the development of the standard.

Use of an IEEE Standard is wholly voluntary. The existence of an IEEE Standard does not imply that there are no other ways to produce, test, measure, purchase, market, or provide other goods and services related to the scope of the IEEE Standard. Furthermore, the viewpoint expressed at the time a standard is approved and issued is subject to change brought about through developments in the state of the art and comments received from users of the standard. Every IEEE Standard is subjected to review at least every five years for revision or reaffirmation. When a document is more than five years old and has not been reaffirmed, it is reasonable to conclude that its contents, although still of some value, do not wholly reflect the present state of the art. Users are cautioned to check to determine that they have the latest edition of any IEEE Standard.

Comments for revision of IEEE Standards are welcome from any interested party, regardless of membership affiliation with IEEE. Suggestions for changes in documents should be in the form of a proposed change of text, together with appropriate supporting comments.

Interpretations: Occasionally questions may arise regarding the meaning of portions of standards as they relate to specific applications. When the need for interpretations is brought to the attention of IEEE, the Institute will initiate action to prepare appropriate responses. Since IEEE Standards represent a consensus of all concerned interests, it is important to ensure that any interpretation has also received the concurrence of a balance of interests. For this reason, IEEE and the members of its societies and Standards Coordinating Committees are not able to provide an instant response to interpretation requests except in those cases where the matter has previously received formal consideration.

Comments on standards and requests for interpretations should be addressed to:

Secretary, IEEE-SA Standards Board
445 Hoes Lane
P.O. Box 1331
Piscataway, NJ 08855-1331
USA

Note: Attention is called to the possibility that implementation of this standard may require use of subject matter covered by patent rights. By publication of this standard, no position is taken with respect to the existence or validity of any patent rights in connection therewith. The IEEE shall not be responsible for identifying patents for which a license may be required by an IEEE standard or for conducting inquiries into the legal validity or scope of those patents that are brought to its attention.

Introduction

(This introduction is not a part of IEEE Std 902-1998, IEEE Guide for Maintenance, Operation, and Safety of Industrial and Commercial Power Systems.)

The purpose of this document is to provide guidelines for the numerous personnel who are responsible for operating industrial and commercial electric power facilities.

The Working Group on a Guide for Operation, Maintenance, and Safety of Industrial and Commercial Power Systems was formed in 1981. It was sponsored by the Industrial and Commercial Power Systems Engineering Committee of the IEEE Industry Applications Society through the Safety, Operations, and Maintenance Subcommittee. The requirements of the then-new Occupational and Safety Health Act (OSHA, a U.S. law) and the limited information that was generally offered at that time were prime driving forces. The first task of the Working Group, a formidable task, was to agree on a scope that would produce a publication of reasonable size. The final product provides basic philosophies and approaches to problems without going into great detail on any one aspect of the subject.

The Working Group recognizes the international applicability of this guide. The Working Group also recognizes that this first edition of the guide refers to some practices that are U.S. oriented. As a practical matter, the consensus was to publish this edition now and to start the first revision promptly, with international content. The Working Group and the Safety, Operations, and Maintenance Subcommittee have committed to incorporating international information in the first revision.

Over the years, a great many people have contributed to the development of this guide. The names of these contributors, to the extent known, are listed below. Undoubtedly, some names have been missed. We extend our apologies to those people for such inadvertent oversight.

At the time this guide was approved, the IEEE Yellow Book Working Group had the following membership:

Erling C. Hesla, *Chair*

Chapter 1: Overview—**H. Landis Floyd II,** *Chair*

Chapter 2: Operating diagrams—**Kenneth W. Carrick,** *Chair*

Chapter 3: System management—**Kenneth W. Carrick,** *Chair*

Chapter 4: System control responsibilities and clearing procedures—**Kenneth W. Carrick,** *Chair*

Chapter 5: Maintenance strategies—**T. John White,** *Chair*

Chapter 6: Maintenance testing overview—**T. John White,** *Chair*

Chapter 7: Introduction to electrical safety—**Joseph J. Andrews,** *Chair*

Chapter 8: Establishing an electrical safety program—**Joseph J. Andrews,** *Chair*

Chapter 9: Providing and maintaining electrically safe facilities—**Joseph J. Andrews,** *Chair*

Chapter 10: Safe electrical work practices—**Joseph J. Andrews,** *Chair*

Chapter 11: Protective equipment, tools, and methods—**H. Landis Floyd II,** *Chair*

Chapter 12: Safe use of electrical equipment—**H. Landis Floyd II,** *Chair*

Jerry S. Baskin
James H. Beall
Carl E. Becker
Richard W. Becker
Kay Bollinger
Thaddeus E. Brown
Barry Brusso
Rene Castenschiold
Paul M. A. Chan
Carey J. Cook
John Cooper
W. H. Cooper
Tim Cotter
John Csomay
James M. Daly
Bruce G. Douglas
Michael J. Foley
Peter J. Gallagher
Edgar O. Galyon
Tom Goavinich
Daniel Goldberg
Terry C. Gould
Dana Hanning Jr.
Raymond N. Hansen
Charles R. Heising
Darin W. Hucul
Howard H. Huffman
Charles Hughes
Robert W. Ingham
R. Gerald Irvine
Gordon S. Johnson
James R. Jones
Ray A. Jones
Robert S. Jordan
W. C. Jordan
Prem P. Khera
Donald O. Koval
Shank T. Lakhavani
C. A. Laplatney
Steven A. Larson
Ralph E. Lee
George E. Lewan
Daniel J. Love
L. Bruce McClung
M. W. Migliaro
John Moore
Robert E. Nabours
Ed Palko
Giuseppe Parise
Elliot Rappaport
Milton D. Robinson
Donald R. Ruthman
H. Kenneth Sacks
Melvin K. Sanders
Vincent Saporita
Lynn F. Saunders
Robert Schuerger
Joe Simon
Robert L. Simpson
Robert L. Smith
Gary Smullin
R. L. Smurif
Stanley Wells
Thomas Wogenrich
Donald W. Zipse

The following persons were on the balloting committee:

Joseph J. Andrews
Arthur Ballato
Jerry S. Baskin
Graydon M. Bauer
James H. Beall
Carl E. Becker
Kenneth W. Carrick
Rene Castenschiold
James M. Daly
H. Landis Floyd II
Jerry M. Frank
Peter J. Gallagher
Daniel Goldberg
James R. Harvey
Erling C. Hesla
Howard H. Huffman
Charles Hughes
Robert W. Ingham
R. Gerald Irvine
Ray A. Jones
Grant C. Keough
Donald O. Koval
L. Bruce McClung
Richard H. McFadden
Ed Palko
James R. Pfafflin
Brian Rener
Milton D. Robinson
Donald R. Ruthman
Melvin K. Sanders
Vincent Saporita
Lynn F. Saunders
Sukanta Sengupta
Conrad R. St. Pierre
T. John White
Donald W. Zipse

The final conditions for approval of this guide were met on 30 October 1998. This guide was conditionally approved by the IEEE-SA Standards Board on 16 September 1998, with the following membership:

National Electrical Code and NEC are both registered trademarks of the National Fire Protection Association, Inc.
National Electrical Safety Code and NESC are both registered trademarks and service marks of the Institute of Electrical and Electronics Engineers, Inc.

Contents

IEEE Guide for Maintenance, Operation, and Safety of Industrial and Commercial Power Systems

Chapter 1
Overview

1.1 Introduction

Even with the best design and equipment, the expected safety and reliability performance of a power system is largely dependent on the quality and capability of its operation and maintenance. Optimizing maintenance and operation often can be the most cost-effective approach in improving system performance.

The phrase "industrial and commercial power systems" covers a broad spectrum. At one end of this spectrum is the large, industrial complex that can justify a staff of highly-skilled and knowledgeable maintenance and operation personnel. At the other end of this spectrum is the small, simple system in which the owner may have little or no electrical expertise.

The objective of this guide is to provide plant engineers with a reference source for the fundamentals of safe and reliable maintenance and operation of industrial and commercial electric power distribution systems. These fundamentals are independent of system size or complexity. The most effective utilization of the information contained in this guide would be its inclusion in a long-term maintenance and operation strategy that is tailored to the individual needs of each power system.

The fundamental elements include

a) Maintenance, operation, and safety considerations in system design;
b) Development of a maintenance and operations strategy to ensure long-term reliability;
c) Development of record-keeping and documentation files;
d) Development and implementation of testing and inspection methods;
e) Development of procedures for auditing maintenance and operation performance;
f) Development of procedures to ensure personnel safety.

1.2 How to use this guide

Chapter 1 provides an overview of this guide.

Chapters 2 through 4 offer guidance for the establishment of administrative procedures, control procedures, and organizational capabilities to ensure a reliable system operation.

Chapters 5 and 6 review various maintenance strategies that are designed to achieve the desired level of performance reliability.

Chapters 7 through 12 provide safety information, a review of available safety equipment, and recommendations on the management of unusual or hazardous activity.

Chapter 2
Operating diagrams

2.1 Introduction

Operating diagrams are the road maps to the operation and maintenance of an industrial or commercial power system. The diagrams can be considered as the primary tool for work on a power system. Without the guidance that is provided by the information contained on these system diagrams, the operation and maintenance of an industrial or commercial complex would be extremely difficult and potentially unsafe.

Vendor's drawings and instruction manuals, plant drawings, maintenance histories, spare parts lists, standard operating procedures, and other documents should be systematically filed and readily available to the personnel who are involved in system operation. A plant's core distribution system may have a life of 30 or more years. Complete, up-to-date equipment documentation is essential not only for supporting routine daily activity, but also for the successful implementation of system expansion or modification. System documentation should include simplified one-line drawings that show only power sources, transformers, voltage levels, major loads, disconnecting devices, switches, and breakers. These drawings are essential for the planning and illustration of operation activity. A master file should be kept in which all revisions are noted and the drawings are subsequently updated because there may be several copies or files of essential drawings that depict the plant. The use of computer-aided drafting systems may expedite the process.

The drawings and the equipment identifications should agree. Frequently, equipment identification is incomplete and poorly located. Staff reliance on such documents may result in errors, if a system is not in place to ensure accuracy. Disconnecting means, switch operators, circuit breaker control switches, circuit breakers, push buttons, motors, and other devices should have complete stand-alone identification nameplates attached in immediate full view (i.e., "FDR 12, Sub. 3" vs. "FDR 12" for a switch control handle). Identification should be unique and affixed to all associated apparatus. The location of the identification nameplate can be as critical as the accuracy of the nameplate.

Critical process operations have been accidentally shut down because circuit breakers in panelboards were not properly identified. The identification of circuits during commissioning should always be verified. A mandatory policy requiring the regular review and updating of panelboard schedules should be instituted.

No standard diagram or type of diagram that is available can suffice for all industrial plants or commercial complexes. No standard electric distribution system is adaptable to all industrial plants because two plants rarely have the same requirements. This chapter outlines the types of diagrams and information that are available to map any distribution system. Consistency is an important factor, however, in keeping diagrams readable and understandable.

2.2 Single-line diagram (one-line diagram)

A reliable single-line diagram of an industrial or commercial electrical power distribution system is an invaluable tool. It is also called a one-line diagram. The single-line diagram indicates, by single lines and standard symbols, the course and component parts of an electric circuit or system of circuits. The symbols that are commonly used in one-line diagrams are defined in IEEE Std 315-1975.[1]

The single-line diagram is a road map of the distribution system that traces the flow of power into and through the system. The single-line drawing identifies the points at which power is, or can be, supplied into the system and at which power should be disconnected in order to clear, or isolate, any portion of the system.

2.2.1 Characteristics of an accurate diagram

The following characteristics should help to ensure accuracy as well as ease of interpretation:

a) *Keep it simple.* A fundamental single-line diagram should be made up of short, straight lines and components, similar to the manner in which a block diagram is drawn. It should be relatively easy to get the overall picture of the whole electrical system. All, or as much as possible, of the system should be kept to one sheet. If the system is very large, and more than one sheet is necessary, then the break should be made at voltage levels or at distribution centers.

b) *Maintain relative geographic relations.* In many cases, it is possible to superimpose a form of the one-line diagram onto the facility plot plan. This is very helpful toward a quick understanding of the location of the system's major components for operating purposes. It may, however, be more difficult to comprehend the overall system operation from this drawing. Such a drawing could be used for relatively simple systems. For more complex systems, however, it should be used in addition to the fundamental single-line diagram.

c) *Maintain the approximate relative positions of components when producing the single-line diagram.* The drawing should be as simple as possible and should be laid out in the same relationship as an operator would view the equipment. The diagram does not need to show geographical relationships at the expense of simplicity.

   NOTE—A site plan with equipment locations may be required to accompany the single-line diagram.

d) *Avoid duplication.* Each symbol, figure, and letter has a definite meaning. The reader should be able to interpret each without any confusion. In this regard, equipment names should be selected before publishing the document; then, these names should be used consistently.

[1]Information on references can be found in 2.4.

e) *Show all known factors.* All details shown on the diagram are important. Some of those important details are as follows:
   - Manufacturers' type designations and ratings of apparatus;
   - Ratios of current and potential transformers and taps to be used on multi-ratio transformers;
   - Connections of power transformer windings;
   - Circuit breaker ratings in volts, amperes, and short-circuit interrupting rating;
   - Switch and fuse ratings in volts, amperes, and short-circuit interrupting rating;
   - Function of relays. Device functions used should be from IEEE Std C37.2-1991;
   - Ratings of motors, generators, and power transformers;
   - Number, size, and type of conductors;
   - Voltage, phases, frequency, and phase rotation of all incoming circuits. The type of supply system (wye or delta, grounded or ungrounded) and the available short-circuit currents should be indicated.

f) *Future plans.* When future plans are known, they should be shown on the diagram or explained by notes.

g) *Other considerations.* Refer to IEEE Std 141-1993 for further discussion of single-line diagrams.

2.2.2 Uses of the single-line diagram

The single-line diagram may be used in a number of important ways in operating and maintaining an industrial or commercial power distribution system. Frequently, the single-line diagram, with all of the listed information, becomes too crowded for information to be used effectively in some of the operating activities. In those instances, it is wise to produce a set of single-line diagrams, with each different diagram in the set containing the pertinent information that is required for a particular activity or set of activities. Some of the needs for special single-line diagrams are

a) *Switching functions.* When the primary use of the diagram is to provide information to system operators for switching in order to isolate portions for maintenance or for load control, then only the data required to make the decisions necessary for system switching are included on the diagram. Sometimes, when the distribution system is complex, a separate version of the single-line diagram in block form is more usable than a complete single-line diagram. This may be identified as a "system operating diagram."

b) *Load flow control.* This diagram is used exclusively for load flow control. It only includes the data that show system component capacities and other data that pertain to load flow.

c) *Relaying and relay logic diagrams.* These single-line diagrams are used to describe the system protective relay systems. These diagrams are used particularly as logic and tripping diagrams that may contain a unique language used only to depict the sequence of relay or system protective component operation under various fault conditions.

d) *Impedance diagram.* This is a single-line diagram that shows the system input impedance and the impedance of all system components in which the impedance of each circuit branch is maintained for system short-circuit analysis. This diagram should include all reactance data on large rotating apparatus or the equivalent data for groups of machines.

2.2.3 Special single-line diagrams

For a large facility that has regularly changing conditions on its electrical distribution system, a mimic bus-type, single-line diagram is very useful. This type of single-line diagram could be a fairly simple pegboard arrangement with movable circuit symbols and tapes in several colors to identify voltage levels; or it could be a light-indicating board that indicates remote switch position via telemeter information and includes load flow data. Modern systems may have this data displayed on a computer monitor.

2.2.3.1 Switching simulator

A switching simulator can be very useful for verifying the effects of any set of switching operations. Low-voltage toggle switches are substituted for circuit breakers and disconnect switches. Signal lamps indicate whether a circuit path has been omitted in the switching sequence, thereby offering the system operator a check on the switching plan prior to performing switching on the distribution system.

Interactive computer programs have been developed to provide the same system modeling as in the electrified-system one-line diagram. These programs can be rapidly manipulated to test any planned switching of an industrial plant's power system. They can be used to check a proposed set of switching instructions and, in some cases, to print a set of instructions to be used in the field.

2.2.4 Schedules

Panel schedules are not one-line diagrams by definition, but they often fulfill the same purpose. Low-voltage lighting and power panel schedules should be maintained as well as all other electrical distribution system diagrams, since they may be used in lieu of the one-line diagram.

2.3 Plan (equipment location plan)

A site plan usually is a necessary accompaniment to the single-line diagram for a complete description and mapping of the industrial and commercial electric distribution system. The locations of the major components of the system are usually easy to visualize; however, circuit routing is difficult to comprehend without a site plan.

Site plans are important for a number of reasons, all of which could affect the operation of the industrial plant or commercial complex at some time. If a major catastrophe such as fire, flood, or storm damage should occur, a site plan would be an important tool if the distribution system were to be reconstructed. The expansion and/or rearrangement of an electrical distribution system could be extremely difficult without accurate records of the location of existing system components. The site plan can be important for identifying the proximity of electrical system components to other maintenance work that may be taking place.

2.4 References

This chapter shall be used in conjunction with the following publications. If the following publications are superseded by an approved revision, the revision shall apply.

IEEE Std 141-1993, IEEE Recommended Practice for Electric Power Distribution for Industrial Plants (*IEEE Red Book*).[2]

IEEE Std 241-1990 (Reaff 1997), IEEE Recommended Practice for Electric Power Systems in Commercial Buildings (*IEEE Gray Book*).

IEEE Std 242-1986 (Reaff 1991), IEEE Recommended Practice for Protection and Coordination of Industrial and Commercial Power Systems (*IEEE Buff Book*).

IEEE Std 315-1975 (Reaff 1993), IEEE Graphic Symbols for Electrical and Electronics Diagrams (Including Reference Designation Letters).

IEEE Std C37.2-1996, IEEE Standard Electrical Power System Device Function Numbers and Contact Designations.

IEEE Std Y32.9-1972 (Reaff 1989), American National Standard Graphics Symbols for Electrical Wiring and Layout Diagrams Used in Architecture and Building Construction.

2.5 Bibliography

Additional information may be found in the following sources:

[B1] *Guide to Industrial Electric Power Distribution*. Compiled by the editors of *Electrified Industry*, Chicago, IL: B. J. Martin Co.

[B2] "The Electric Power System," *Plant Engineering*, Oct. 15, 1981.

[2]IEEE publications are available from the Institute of Electrical and Electronics Engineers, 445 Hoes Lane, P.O. Box 1331, Piscataway, NJ 08855-1331, USA (http://www.standards.ieee.org/).

Chapter 3
System management

3.1 Introduction

A well designed and constructed power system will not provide a safe and reliable operation unless it is properly managed. Any electrical power distribution system, from the smallest system to the largest and most complex system, needs to be managed. As systems become larger in size and complexity, the problems of system management increase, thereby requiring more time and attention from the system-operating personnel.

Good design, proper installation, quality assurance, and sound operating and maintenance programs provide the basic foundation for the safe and reliable operation of industrial electric power systems. A plant engineer who is faced with the task of improving the plant's electric power system performance, however, will likely find that programs to reduce human error are more cost-effective than system modifications or additional preventive maintenance. In fact, given good design and a sound maintenance program, the inherent system reliability can only be achieved by the reduction of operating error.

The operation of an electric power system should also address the problem of human errors. The following examples should be considered:

- Following a severe thunderstorm, a plant shift supervisor made a walk-through inspection of the plant's primary distribution switchgear. Upon seeing a red light for each circuit breaker, he immediately tripped each circuit breaker in order to obtain a green-light indication. Because he incorrectly thought that the red light meant "open," he shut down the entire plant.
- One of a plant's two steam boilers was down for annual inspection and maintenance. An electrician who was assigned to make a modification to the boiler control circuit erroneously began working on the operating boiler control circuit and shut down the operating boiler.
- An investigation of a 15 kV outdoor bus duct fault revealed that production personnel routinely turned off outside lighting at the beginning of the day shift by switching off circuit breakers in a 120 V distribution panel. The bus duct heater circuit was incorrectly identified, and was being switched off with the lighting circuits.

It is a natural tendency to blame equipment for failures, rather than human error. The bus duct fault in the last example could have been classified as an equipment failure; however, the prime cause was improper operation (human error) of the bus duct heaters.

Most plant electrical outages that clearly are not due to equipment failure, lightning, or utility disturbances can be prevented by making an objective investigation of the potential for outages and by following these guidelines:

a) Document the system and identify the equipment.
b) Plan switching operations in detail.
c) Secure equipment from unintentional operation.
d) Clearly define operating responsibility and adhere to it rigidly. System operation can and should be managed.

Effective managers of a power system will consider load distribution, system integrity, power factor, system protection coordination, and operating economics. Each of these areas is discussed in this chapter, thus showing how all of these considerations relate to each other. No area of industrial and commercial power system management is independent of the other.

3.2 Load distribution

How and where loads are connected to a distribution system normally are determined early in the system design. If the logic that is used in determining the load arrangement has been documented by the system designers, then the system operators may obtain and understand that logic. If it has not been documented, the logic should be developed later. Information for the development of that logic can be obtained from interviews with the system designers, if they are available, or by studying the loads and classifying them by type. Loads should be classified by their criticality to the operation of the facility that is served by the electrical distribution system. For example, loads such as boiler fans or boiler feedwater pumps would normally be listed as more critical to an industrial operation than the load of a single production area because loss of the boiler area could trigger a chain reaction that would affect the entire industrial plant.

The importance of various loads should be kept foremost in the system operator's mind when planning or performing any switching of the system. Switching should be done in such a manner that the integrity of service to critical loads is maintained or, at least, the possible increased exposure to service interruption is minimized. The system operator should have studied the consequences of a service interruption, and what actions to take should an interruption occur. Because the exact nature or cause of a service interruption cannot be determined before the fact, a definite course of action cannot be predetermined. General guidelines can be established, however, such as protecting the integrity of alternate sources. It is important not to use all alternatives and therefore get the power system extended into a long series of circuits such that a single failure can cause a total system collapse. Some computer programs can assist greatly in determining the consequences of each switching action.

The system operator should always monitor the electrical system load distribution in terms of nominal electrical load measurement parameters such as watts, vars, and amperes to ensure that some circuits are not overloaded while other circuits are underutilized. Where parallel or alternate circuits are available to carry the load, the system load should be balanced between the circuits, if the system connections make it possible. Critical service loads should be served from alternate circuits so that a single outage on one of the alternate circuits will not remove service from all critical loads. This concept may create a situation in which the load magnitudes are not balanced. The judgment should be made as to whether the operator wants load balance or a reduced probability of the upset of critical facility operations.

3.3 System integrity

Operation of electrical distribution systems should keep the system whole (totally in service) for as much of the time as possible. When a system has redundant circuits, as in the case of primary or secondary selective systems, the amount of time of operation as a radial system should be held to a minimum when an alternate circuit is removed from service for maintenance or other reasons. This is not intended to restrict maintenance or repair time, but to ensure that system integrity is maintained while a circuit is out of service. (See IEEE Std 141-1993.)[1]

A complex industrial plant may contain many redundant features that are related to the electrical power distribution systems. Redundancy allows for maintenance or repair on a portion of the system with minimum disruption to plant production. Some traps exist, however, that can easily undermine the reliability of redundant systems. Subclauses 3.3.1 through 3.3.4 discuss some of these traps and ways in which they can be avoided.

3.3.1 Considering outside forces

All external influences should be considered when protecting the integrity of the electrical distribution system. The system should be protected, to the greatest extent that is practical, from damage or interference from outside sources. The operator needs to protect the system from negative environmental influences. The abbreviated list below outlines some of the factors for consideration.

a) Maintain good housekeeping at all times. Good housekeeping in the substations and around all apparatus is necessary if uninterrupted service is to be maintained.
b) Strictly avoid using electrical rooms and spaces for manufacturing or storage, except for minor parts that are essential to the maintenance of the installed equipment.
c) Provide good general maintenance consistently. This applies to area maintenance as well as to electrical maintenance.
d) Carefully consider the possible need for operation of cranes in the area where outdoor bare conductors are used as a part of the distribution system. The unexpected movement of crane booms near energized power lines can affect system integrity as well as the safety of personnel in the area.

3.3.2 Equipment location

Ideally, distribution equipment should be isolated in a locked area, either indoors or outdoors, and should be accessible only to qualified personnel. Some electromechanical protective relays are vulnerable to vibration and accidental jarring, which could cause undesired relay operations. Janitors' brooms have an uncanny ability to find a control switch or bump the most sensitive relay. If substations or distribution switching equipment are located near pedestrian or vehicular traffic ways, traffic paths should be clearly marked and identified to keep personnel and vehicles away from the electrical equipment.

---

[1]Information on references can be found in 3.7.

3.3.3 Congested construction/maintenance activity

During maintenance shutdowns or during construction tie-in activity, the area around distribution equipment can become congested with materials and personnel. The protection of nearby operating equipment from accidental operation should be addressed. In redundant systems, such as primary selective or secondary selective distribution arrangements, a portion of the system may be energized while the equipment adjacent to it is serviced. In this situation, the two pieces of equipment should be individually barricaded, with personnel access strictly controlled during the periods of maintenance or construction. No exposure should exist that would allow personnel to accidentally bump against a control switch or protective relay.

3.3.4 Operating integrity

Operating integrity of the power system will also be affected adversely by inadequate system maintenance programs. Failure to perform maintenance work on a system adversely affects its integrity. The system operator should evaluate the merits of a redundant system, with portions that are occasionally out of service for maintenance, vs. waiting for a total shutdown on a simple radial system. The best maintenance plan for the particular power system in question should be determined.

3.4 Power factor

Low power factor will reduce system capacity. This reduction occurs because the equipment, particularly transformers and wiring, is forced to carry larger currents than would be necessary otherwise. This results in the increased heating of the equipment and conductors, as well as an increased voltage drop on the distribution circuits. Because of reduced voltage, utilization equipment operates less efficiently and motors overheat. The effects of low power factor are manifest throughout the distribution system back to, and including, the source (e.g., a utility company tie or self-generation). Some utility tariffs include power-factor penalty clauses that add surcharges on the utility bill if a facility does not keep its power factor above a predetermined value. The following should be noted:

a) System operators need to be aware of low power factor anywhere on the power system so that they can evaluate and correct the situation.
b) Correction can take the form of adding power-factor correction capacitors either to motor circuits or as shunt banks for system sections or entire systems.
c) The operator should take advantage of any synchronous machines that may be on the system by using them to supply reactive flow into the system. Typically, synchronous generators are operated with a leading power factor to supply vars to the system. Synchronous motors usually are not loaded to their nameplate capability so these motors can be operated in an overexcited condition to supply vars to the power system.
d) The use of high-power-factor lighting ballasts will avoid introduction of power-factor problems, and may improve the plant power factor significantly if lighting is a substantial part of the system load.

3.5 System protection coordination

When an electrical distribution system is designed and constructed, a fault-current coordination study should be conducted, and circuit protective devices should be sized and set according to the results of the study. In time, however, the electrical system configurations are often changed due to the changing needs of the end users. If the coordination and capability of the electrical equipment are not reviewed at the time of the changes, faults could result in unnecessary tripping of a main breaker or, even worse, an explosion of equipment that was thought to be in good condition. When system conditions change, the results that were obtained in the original fault-current coordination study may no longer apply to the current system. Unnecessary tripping, known as lack of selectivity, could be caused by poor coordination. An equipment explosion could result from the interrupting capability of the circuit breaker being exceeded. Both indicate a clear need for an updated fault-current coordination study.

3.5.1 Utility systems delivering higher fault currents

The demand for electricity, particularly in the industrial and commercial environment, has been steadily increasing. Consequently, utility systems have grown much larger and have become capable of delivering much higher fault-currents at service points than in the past. Therefore, protective devices that were properly applied at the time they were installed may have become inadequate after system changes, and the protective system may no longer be coordinated. When available fault current increases to the point at which it exceeds protective device interrupting and withstand ratings, violent failure is possible, regardless of how well the devices are maintained.

3.5.2 Protection in an electrical distribution system

System and equipment protective devices are a form of insurance. This insurance pays nothing as long as there is no fault or other emergency. When a fault occurs, however, properly applied protective devices reduce the extent and duration of the interruption, thereby reducing the exposure to personal injury and property damage. If, however, the protective system does not match system needs, just as an insurance policy should keep up with inflation, it is no help at all. It is the responsibility of the system operator to ensure proper system protection and coordination.

3.5.3 Protective equipment set to sense and remove short circuits

In medium-voltage systems, the protective equipment for feeder conductors is often set to sense and remove short circuits, but not necessarily to provide overload protection of circuits. Device settings sometimes are purposely chosen low enough to sense and provide a degree of overload protection. Operators should be aware of this so that a protective device that is set lower than necessary for coordination does not cause a false tripping action during system switching procedures. System protection coordination is an important consideration in switching systems with loop feeds and alternate sources. To avoid false tripping action, operators should be aware of the settings and any probable temporary overloads or circulating currents during switching.

3.6 Operating economics

It is important to operate an electrical distribution system economically because of the high costs of losses and the cost of system expansion. Today, there are numerous methods for monitoring and controlling the power flow through the distribution system. These methods range from simple ammeter, voltmeter, wattmeter, and varmeter systems to complex supervisory control and data acquisition systems. A system can be designed to fit the needs and budget of any size facility.

3.6.1 Energy conservation

Energy conservation is the key to the economic operation of a power system, regardless of the methods that are used to monitor and control the energy flow through the system. Energy conservation begins with thorough and complete design practices. The system should be operated in such a manner as to keep losses to a minimum and to minimize any utility power factor or demand charges.

3.6.2 Power-factor correction

Power-factor correction, by the addition of capacitors at the facility service point, reduces power-factor charges from the serving utility. This, however, does not release any capacity of the load-side distribution system. Power-factor correction, closer to the loads, reduces currents in the main feeder conductors. This reduces the system losses, reduces power-factor billing charges, releases circuit capacity, and improves voltage regulation. The release of circuit capacity may be used to avoid costly system expansion projects by allowing additional circuit loading.

3.6.3 Utility demand charge

Most utilities have a demand charge that is based on kilovolt-amperes and kilowatts, or kilovolt-ampere-hours and kilowatt-hours, which automatically includes power factor, and they charge a financial penalty for loads that operate below a specified minimum power factor. The demand level is dependent upon the type of industrial plant or commercial facility. The system operator should develop the logic of that operation so that effective demand control can be practiced. Demand charges normally are maintained at peak levels for finite time periods after a new peak is established. The cost of a single peaking event could have a recurring cost for as long as 12 months. Lack of demand control can escalate one apparently small indiscretion into a very expensive event. The unnecessary operation of spare equipment that adds load to the system, even for a short time, should be avoided so as not to increase demand peaks. The operator should be aware of the serving utility rate/demand structure in order to operate at peak effectiveness and to avoid any unnecessary demand charges.

3.7 References

This chapter shall be used in conjunction with the following publications. If the following publications are superseded by an approved revision, the revision shall apply.

IEEE Std 141-1993, IEEE Recommended Practice for Electric Power Distribution for Industrial Plants (*IEEE Red Book*).[2]

IEEE Std 241-1990 (Reaff 1997), IEEE Recommended Practice for Electric Power Systems in Commercial Buildings (*IEEE Gray Book*).

IEEE Std 242-1986 (Reaff 1991), IEEE Recommended Practice for Protection and Coordination of Industrial and Commercial Power Systems (*IEEE Buff Book*).

3.8 Bibliography

Additional information may be found in the following sources:

[B1] Beeman, D., ed., *Industrial Power Systems Handbook.* New York: McGraw-Hill, 1955.

[B2] Smerton, R. W., ed., *Switchgear and Control Handbook.* New York: McGraw-Hill, 1977.

[B3] "The Electric Power System," *Plant Engineering*, Oct. 15, 1981.

[2]IEEE publications are available from the Institute of Electrical and Electronics Engineers, 445 Hoes Lane, P.O. Box 1331, Piscataway, NJ 08855-1331, USA (http://www.standards.ieee.org/).

Chapter 4
System control responsibilities and clearing procedures

4.1 Introduction

Everyone who interacts with the power supply and distribution system of an industrial or commercial establishment has some responsibility for the control of that system. The owner has primary and ultimate responsibility for all operating occurrences on the power distribution system. Those who perform maintenance on the system, any contractors or others who may work on the system, and the utility, however, also have control responsibilities.

The complexity of control, and the level of technical and management attention required, increases with the complexity of the system. A simple, single-source radial supply system with one distribution panelboard or switchboard usually does not need detailed clearing procedures (i.e., de-energizing a portion of the system and/or shifting load) for making the system ready for work. A system with loop feeds and alternate sources requires carefully studied and detailed procedures to ensure personnel safety.

4.2 Responsibility of the owner

The owner has the ultimate responsibility for system control. This includes the responsibility of being assured that all other parties perform at a level that is consistent with their responsibilities. In effect, the owner becomes responsible, to some degree, for the actions of all parties that perform work on or with the power distribution system in the facility.

The owner, or the owner's representative(s), should be aware and knowledgeable of all work that is performed on the system. This does not mean that the owner shall possess the technical details of any and all work to be performed; however, a general understanding is required. The owner needs to provide for the de-energization of the system, isolation, lockout and/or tagout, and testing that are required before personnel are allowed to work on any portion of the system.

If the system is large and complex, the owner or the owner's representative(s) normally should utilize their clearing procedures to get the system ready. This usually includes all switching that is required to de-energize the system or portion of the system in the work area that applies grounding devices and the necessary tagging or locking-out of all the isolating devices. On a simple radial feeder system, personnel may be permitted to open the isolating device lock and/or tag the isolating device themselves and apply grounding devices when necessary. The owner still is responsible to see that it is done properly.

Another responsibility of the owner is to control access as required by law. The owner is responsible for the protection of the general population of the location from possible harm by making certain portions or components of the power distribution system accessible only to qualified persons. That responsibility can be carried out, in part, by the proper posting of warning signs.

4.3 Maintenance role

Maintenance personnel shall take the necessary steps to ensure that the system or portion of the system on which they plan to work is in an electrically safe working condition. They should then add any additional protective devices, such as locks and tags. Before maintenance personnel start any work on the power distribution system, they are responsible for assuring themselves that the system is safe for work. Maintenance personnel should maintain the system isolation devices and lockout mechanisms in good working order. For a more complete discussion concerning safe practices and lock and tag requirements, see 10.4.2.

4.4 Utility responsibilities

It is the responsibility of the serving utility company to switch for isolation of the utility power supply prior to the service point to any facility. The utility shall supply to the facility owner or the owner's representative(s) a point at which to padlock the service disconnect or provide some other means of service isolation control.

It is also the primary responsibility of the serving utility company to protect the service from possible disturbances or damage from outside influences. The serving utility shall also protect its service equipment from access by unauthorized personnel.

4.5 Other workers

Outside contractors (i.e., nonowner personnel) shall assure themselves that they have an understanding of the system on which they are going to work and how it is isolated from all energy sources. They should understand how isolation and protection are achieved, and should comply with the owner's clearance procedures to ensure their continued safety. Contractor employees should add their own lockout/tagout device(s) to the appropriate isolation point(s). Relevant drawings and other information should be made available and should be reviewed with persons working on the system. (See Chapter 2 for a more complete discussion of drawings.)

4.6 Clearing procedures

The complexity of the system normally determines the level of detail planning that is required for system clearing procedures. A simple, single-source, radial supply system may only require opening a single switch or circuit breaker for circuit isolation. The clearing procedures for even so simple a case, however, should include checking to ensure that no other sources exist and that the correct isolating device is being operated. It is important that all persons who may be exposed to a hazard, as a result of a switching action, be notified prior to the action.

Complex power distribution systems that require several switching steps to isolate a portion of the system require more elaborate clearing procedures. It is necessary to use written switching instructions for systems that may have several sources into an area. When written instructions are used, a third party, who is familiar with the power system, should review them for errors and omissions. The consequences of learning about switching errors while in the act of switching are usually costly, especially when the wrong portion of the system is accidentally de-energized. It is important that written procedures be shared with all persons who are involved in the switching process.

A single-line diagram should accompany the written switching instructions so that the switch operator can keep track of the progress through the system. A real-time, single-line mimic bus on a very complex system allows for the independent monitoring of the switching process through the system as component status is changed. Some mimic-bus systems allow the operator to simulate switching of the system off-line, which allows for the detection of possible errors before the actual switching is performed.

The clearing procedures should be completely written, checked, and understood by all persons involved before they are applied to any portion of the power distribution system. The instructions and/or procedures should include a verification that the power has been removed (by live-line testing or other means) followed by the placement of grounds and the locking/tagging of isolating devices.

4.7 References

This chapter shall be used in conjunction with the following publications. If the following publications are superseded by an approved revision, the revision shall apply.

Accredited Standards Committee C2-1997, National Electrical Safety Code® (NESC®).[1]

IEEE Std 141-1993, IEEE Recommended Practice for Electric Power Distribution for Industrial Plants (*IEEE Red Book*).[2]

IEEE Std 241-1990 (Reaff 1997), IEEE Recommended Practice for Electric Power Systems in Commercial Buildings (*IEEE Gray Book*).

NFPA 70E-1995, Standard for Electrical Safety Requirements for Employee Workplaces.[3]

[1]The NESC is available from the Institute of Electrical and Electronics Engineers, 445 Hoes Lane, P.O. Box 1331, Piscataway, NJ 08855-1331, USA (http://www.standards.ieee.org/).

[2]IEEE publications are available from the Institute of Electrical and Electronics Engineers, 445 Hoes Lane, P.O. Box 1331, Piscataway, NJ 08855-1331, USA (http://www.standards.ieee.org/).

[3]NFPA publications are available from Publications Sales, National Fire Protection Association, 1 Batterymarch Park, P.O. Box 9101, Quincy, MA 02269-9101, USA (http://www.nfpa.org/).

Chapter 5
Maintenance strategies

5.1 Introduction

The research and analysis of the optimum maintenance program for electrical equipment have been ongoing tasks of the maintenance manager for as long as electrical equipment has been used to support facility or plant operation. In spite of the findings from decades of analysis, maintenance programs still vary from breakdown maintenance programs to very sophisticated preventive maintenance programs. Preventive maintenance programs may include predictive maintenance as well as a more sophisticated reliability-centered program. The many variables that exist, from the types of electrical equipment to the types of applications in which they are used, make the universal definition of an exact maintenance program very difficult. It is believed almost universally, however, that some form of maintenance is necessary.

NFPA 70B-1994 [B3][1] states the following:

"Electrical equipment deterioration is normal, but equipment failure is not inevitable. As soon as new equipment is installed, a process of normal deterioration begins. Unchecked, the deterioration process can cause malfunction or an electrical failure."

It is necessary to control equipment deterioration in order to maintain the use for which the equipment and systems were originally designed and installed. Although most parties would agree that preventive maintenance is necessary in order to ensure the reliability of electrical power systems, there remains a wide disparity as to the content of a preventive maintenance program.

5.2 Definitions and acronyms

5.2.1 Definitions

For the purposes of this guide, the following terms and definitions apply. IEEE Std 100-1996[2] should be referenced for terms not defined in this subclause.

5.2.1.1 breakdown maintenance: Those repair actions that are conducted after a failure in order to restore equipment or systems to an operational condition.

5.2.1.2 electrical equipment: A general term that is applied to materials, fittings, devices, fixtures, and apparatus that are a part of, or are used in connection with, an electrical installation. This includes the electrical power generating system; substations; distribution systems

[1]The numbers in brackets correspond to those of the bibliography in 5.7.

[2]Information on references can be found in 5.6.

including cable and wiring; utilization equipment; and associated control, protective, and monitoring devices.

5.2.1.3 maintenance: The act of preserving or keeping in existence those conditions that are necessary in order for equipment to operate as it was originally intended.

5.2.1.4 predictive maintenance: The practice of conducting diagnostic tests and inspections during normal equipment operations in order to detect incipient weaknesses or impending failures.

5.2.1.5 preventive maintenance: The practice of conducting routine inspections, tests, and servicing so that impending troubles can be detected and reduced or eliminated.

5.2.1.6 reliability-centered maintenance (RCM): A systematic methodology that establishes initial preventive maintenance requirements or optimizes existing preventive maintenance requirements for equipment based upon the consequences of equipment failure. The failure consequences are determined by the application of the equipment in an operating system.

5.2.2 Acronyms

EPM electrical preventive maintenance
NETA InterNational Electrical Testing Association
RCM reliability-centered maintenance
UPS uninterruptible power supply

5.3 Preventive maintenance

5.3.1 Philosophy

Most people recognize the need for the maintenance of electrical equipment. The debate really focuses on how much maintenance is enough. The key to the discussion over the proper amount of maintenance centers on the economic balance between the cost of performing maintenance and the importance of reliable power. For example, a computer center with a downtime cost of $100 000 or more an hour would justify a much more extensive maintenance program than would a small facility whose downtime cost might be minuscule in comparison.

Moreover, it has been shown that there is a balance to the amount of economic benefit that is achieved from performing maintenance. A lack of maintenance eventually results in failures and a high cost to a plant. Likewise, an extreme amount of maintenance is wasteful and also results in a high cost to a plant. The optimum maintenance program lies somewhere in between. This balance point can vary for different types of facilities.

There are two benefits to having an effective preventive maintenance program. The first is that costs are reduced through the minimizing of equipment downtime. The second benefit is obtained through improved safety and system performance. Other intangible benefits include things such as improved employee morale, better workmanship, increased productivity, reduced absenteeism, reduced interruption of production, and improved insurance considerations. In planning an electrical preventive maintenance (EPM) program, consideration must be given to the costs of safety, the costs associated with direct losses due to equipment damage, and the indirect costs associated with downtime or lost or inefficient production.

5.3.2 Design considerations

The best preventive maintenance programs start during the design of the facility. A key design consideration in order to support preventive maintenance is to accommodate planned power outages so that maintenance activities can proceed. For example, if delivery of power is not a 24 hour necessity, then extended outages after normal work hours can be allowed for maintenance activities. Otherwise, consider design features that can speed up the maintenance process or reduce the duration of the outage to loads. These might include redundant circuits, alternate power sources, or protective devices such as drawout circuit breakers (rather than fixed-mount circuit breakers).

Additional consideration should be given to the accessibility of the electrical equipment for maintenance. Circuit breaker location can be critical to the maintenance process. An example would be circuit breakers that are installed in a basement that has only stairway access through which equipment can be brought down to the circuit breaker location. In addition, access to the back of switchboards or switchgear, as opposed to their being mounted against the wall, may be necessary in order to perform thorough maintenance.

The environment in which the equipment is installed can play an important part in maintenance. Where equipment is mounted (inside or outside) and whether it is properly enclosed and protected from dust, moisture, and chemical contamination are all factors that influence the frequency with which maintenance tasks should be performed.

The design phase is also the period in which the establishment of baseline data for the equipment should be considered. This can be done by including in the design specifications the acceptance or start-up testing of the equipment when it is first installed. The InterNational Electrical Testing Association (NETA) provides detailed specifications for electrical power equipment in NETA ATS-1995 [B1].

Design drawings are very important to an effective maintenance program. As-built drawings should be kept up-to-date. An accurate single-line diagram is crucial to the effective and safe operation of the equipment. This helps the operator to understand the consequences of switching a circuit that can interrupt power in an undesirable or unplanned mode. More significantly, it can help avoid the accidental energization of equipment.

As part of the procurement of the electrical equipment, consideration should be given to the tools and instruments that are required to perform effective maintenance, such as hoists or manual-lift trucks that are used to remove and install circuit breakers. These tools and instruments will help to ensure safety and productivity. Finally, the installation, operation, and maintenance manuals should be obtained and filed.

5.3.3 Creating an EPM program

To be successful, a preventive maintenance program shall have the backing of management. There should be the belief that operating profit is increased through the judicious spending of maintenance dollars. Financial issues should be considered when evaluating the need for continuous electrical power. These factors will help to dictate the level of importance that a facility places on a preventive maintenance program. The cost of downtime or lost production, and how that can be minimized through effective maintenance, also should be considered.

A complete survey of the plant should be performed. This survey should include a listing of all electrical equipment and systems. The equipment should be listed in a prioritized fashion in order to distinguish those systems or pieces of equipment that are most critical to the operation. The survey should also include a review of the status of drawings, manuals, maintenance logs, safety and operating procedures, and training and other appropriate records. It should be recognized that the survey itself can be a formidable task. It is likely that power outages may be required in order to complete the survey. The gathering of documentation is important. This includes not only the drawings of the facilities, but also all the documentation that is normally provided by the manufacturer of the equipment. The manufacturer's manuals should include recommended maintenance procedures, wiring diagrams, bills of materials, assembly and operating instructions, and troubleshooting recommendations.

Next, the necessary procedures for maintaining each item on the list should be developed. NFPA 70B-1994 [B3] and NETA MTS-1993 [B2] are valuable resources that provide much of this information. Procedures should also be developed that integrate the equipment into systems. People that are capable of performing the procedures should be selected and trained. At some level of technical performance, it may be desirable to contract parts of the maintenance program to qualified outside firms, particularly those functions that require special test equipment to perform.

Finally, a process shall be developed to administer the program. This process may be manual or software-based. There are many commercially available systems with varying levels of sophistication.

Consideration also shall be given to some of the less technical parts of the process. Pre-maintenance considerations might include the logistics of getting equipment in and out of the area to be maintained, general safety procedures, procedures to be followed in the event of an emergency, and record-keeping that has to be accomplished ahead of the maintenance activity, as well as follow-up maintenance, special lighting needs, and equipment-specific safety precautions. In addition, an ongoing task is that of keeping access to electrical equipment free from being blocked by stored materials, such as spare parts. Record keeping and maintenance follow-up activities also shall be considered.

5.3.4 Reliability-centered maintenance (RCM)

RCM is a systematic and comprehensive approach to the maintenance of systems and equipment in which reliability is critical. RCM was initially developed by the airline industry, but has since been adopted by several other industries. The process involves analyzing the criticality and failure mode, and then determining for each component what is the appropriate and effective preventive maintenance activity. All possible failure modes and appropriate preventive maintenance are evaluated for each component, thus establishing a program that is proven to result in improved facility reliability. Other benefits include improved safety, decreased repair costs, shorter outages, reduction in overhaul frequency, and improved management of spare parts.

The following outlines the important philosophical topics surrounding RCM:

a) RCM is simply organized, documented, common sense.
b) The focus of RCM is to analyze and make changes to current preventive maintenance program activities in order to improve their applicability and cost-effectiveness.
c) RCM evaluations determine the maintenance requirements of equipment based upon the function of the system and how the equipment supports the function. An RCM-based maintenance program provides a technical basis for preventive maintenance activity that is not strictly based upon regulatory requirements, vendor recommendations, or industry standards.
d) RCM focuses preventive maintenance resources on equipment that is critical to maintaining important system functions. RCM emphasizes proactive, predictive maintenance techniques over traditional, time-directed maintenance techniques.

Implementing an RCM program involves taking an existing EPM program and modifying it to align with the "critical path" or "mission" of the facility. In a data center, that would include everything that is necessary to keep the computers running, which includes the uninterruptible power supply (UPS) as well as the air conditioning system. All the systems in the critical path would receive priority, while the remaining nonessential support systems would receive much less attention.

5.3.5 Specialized equipment

The maintenance of electrical power systems requires a wide range of materials, tools, and test equipment. Normal hand tools and maintenance supplies, such as cleaning materials, vacuums, and other shop-oriented tools, are necessary. At the other end of the spectrum are specialized test instruments, such as high-current test sets that are necessary for primary injection of low-voltage circuit breakers. Some protective devices now require computers in order to perform calibration.

Safety equipment is also necessary for maintenance. This may include safety grounds, flame-retardant clothing, and face shields; protective rubber goods such as blankets, gloves, and sleeves; and insulated hand tools such as screwdrivers and pliers. Also needed for safety are portable meters or instruments to determine the presence or absence of voltage. This determination is necessary at each voltage level within the facility.

5.3.6 Record keeping

Each maintenance interval should be documented by listing the equipment that is being serviced and the maintenance procedures that are being applied. Data sheets that record the results of all tests should be completed.

These data should be analyzed by maintenance management in order to help determine the condition of the equipment and to determine the necessary repairs. Repairs could be categorized by their critical nature, such as whether the repairs need to be done immediately before returning to service or whether they could be scheduled for a future date. The results of the maintenance records and the analysis of the data should lead management toward either increasing or decreasing the maintenance frequency, or toward determining that the frequency and procedures are adequate.

A short-circuit and coordination study should be maintained up-to-date. This document not only helps plant maintenance personnel evaluate the suitability of protective devices for safely interrupting fault currents that can occur in the plant, but also gives the information that is necessary to set adjustable circuit breakers and protective relays to provide for the optimum selectivity. Proper coordination will minimize plant power outages in the event of equipment failure.

5.4 Fundamentals of electrical equipment maintenance

5.4.1 Inspection and testing

The condition of electrical equipment is generally affected by the atmosphere and conditions under which the equipment is operated and maintained. Water, dust, temperature, humidity, corrosive fumes, vibration, and other environmental factors can adversely affect electrical equipment. Electrical equipment life can be extended dramatically by simple precautions that promote cleanliness, dryness, tightness, and the prevention of friction.

The thoroughness of maintenance procedures can be categorized into three different levels:

Level 1—General inspection and routine maintenance;

Level 2—Inspection, general tests, and preventive maintenance;

Level 3—Inspection, specific tests, and predictive maintenance.

Testing would include

a) Insulation tests;
b) Protective device tests;
c) Analytical tests (e.g., time travel analysis, dissolved gas analysis, infrared, and contact resistance);
d) Grounding tests;
e) Functional tests.

Chapter 10 describes each of these tests in detail.

The following equipment should be in the maintenance program:

— Switchboards and switchgear assemblies
— Disconnecting switches
— Circuit breakers
— Capacitors
— Surge arresters
— Current transformers
— Voltage transformers
— Protective relays
— Network protectors
— Fuses
— Batteries and battery chargers
— Meters and other instruments
— Alarms and alarm systems
— Grounding
— Ground detection schemes
— Transformers
— Insulating liquids
— Cables
— Busways
— Busducts
— Motor control center and motor starters
— Motor protective devices
— Motor drives
— Transformer auxiliary systems
— Rotating equipment
— Lighting
— Wiring devices
— Uninterruptible power supplies
— Transfer switches
— Test and safety equipment

5.4.2 Repairs

Repairs can be categorized by their sense of urgency. Some repairs must be accomplished before the equipment can be returned to service. Other repairs may require material items that are not stocked, and cannot be accomplished until those items have been received and properly installed. Some repairs can be postponed, thus allowing the electrical system to go back into service without undue risk. In this method, the repair could be scheduled for a future date when it is more convenient to the plant.

A part of EPM is determining which spare equipment or parts should be kept in stock, such as fuses, circuit breakers, and other components, in order to be able to repair critical items and return a shut-down facility to operation. This, like the maintenance procedure itself, is an economic benefit vs. cost of inventory balancing act.

5.4.3 Failure analysis

When equipment fails, it is important to understand the reason why. Failure analysis, when done properly, locates the root cause of the failure. This is important in order to take the necessary steps to prevent similar failures in the future. Failure analysis involves an effort to reconstruct, at least mentally, the conditions that existed prior to failure and the events that led to the nature of the failure. It is through this process that the root cause can be determined.

There are engineers that specialize in forensic and failure analysis. These people, through their experience, are generally able to recognize failure patterns and to draw accurate conclusions much more readily than the untrained person. This is a specialty that is generally contracted when firms do not have that capability in-house.

5.5 Inspection and test frequency

Equipment in a critical service would generally receive maintenance attention more frequently than other equipment. Manufacturer's service manuals should be consulted in determining an adequate frequency. They generally give frequencies that are based upon a standard, or average, or upon operating conditions. This is a good basis from which to start in determining the frequency for a given facility. A good guide for both maintenance frequency and routine inspections and tests is found in NFPA 70B-1994 [B3]. Proper maintenance record keeping, together with periodic reviews, should reveal where adjustments to the frequency may be necessary, based on the actual effectiveness of the maintenance program.

5.6 Reference

This chapter shall be used in conjunction with the following publication. If the following publication is superseded by an approved revision, the revision shall apply.

IEEE Std 100-1996, IEEE Standard Dictionary of Electrical and Electronics Terms.[3]

5.7 Bibliography

Additional information may be found in the following sources:

[B1] NETA ATS-1995, Acceptance Testing Specification.

[B2] NETA MTS-1993, Maintenance Testing Specification.

[B3] NFPA 70B-1998, Recommended Practice for Electrical Equipment Maintenance.[4]

---

[3]IEEE publications are available from the Institute of Electrical and Electronics Engineers, 445 Hoes Lane, P.O. Box 1331, Piscataway, NJ 08855-1331, USA (http://www.standards.ieee.org/).

[4]NFPA publications are published by the National Fire Protection Association, Batterymarch Park, Quincy, MA 02269, USA (http://www.nfpa.org/). Copies are also available from the Sales Department, American National Standards Institute, 11 West 42nd Street, 13th Floor, New York, NY 10036, USA (http://www.ansi.org/).

Chapter 6
Maintenance testing overview

6.1 Introduction

Maintenance testing is an important procedure that is used to detect deficiencies in electrical equipment before the equipment fails in service, often catastrophically. The very nature of electricity creates this need for testing. In normal operation, electrical equipment transports and utilizes energy. When a problem arises, such as an overload or a short circuit, the system should detect the problem and isolate it in a safe manner. The only way to know if the protective system works before it is needed is through testing. By simulating the various failure modes with nondestructive test methods, deficiencies in the system can be located and corrected. Then when the system is called upon to operate in a fault condition, testing helps to ensure that it does its job properly and safely.

Electrical equipment is used in a variety of categories with respect to energy. Power flow starts with generation equipment. Next, electrical equipment such as transmission and distribution lines, power cables, and bus structures are used to transport energy. Other equipment, such as transformers, rectifiers, and converters, will change the form of the energy in some manner. A fourth category of equipment is used to detect and isolate problems. This includes the sensing devices, such as protective relays, along with the interrupting devices, such as circuit breakers, switches, or fuses. The final category of equipment utilizes the energy to perform work in such forms as motors, solenoids, or electric heaters.

Maintenance testing needs may vary with each of these categories. Keep in mind that the purpose of maintenance testing is to determine if the equipment will continue to properly perform its function. In many cases, the testing consists of simulating different operating conditions and evaluating how the equipment responds.

6.1.1 Acronyms and abbreviations

CT current transformer
DLA dielectric loss angle
hi-pot high potential
PT potential transformer
RCM reliability-centered maintenance
TTR transformer turns ratio
VT voltage transformer

6.2 Insulation tests

One characteristic that all types of electrical equipment have in common is the use of some form of insulation. At its most basic level, all electrical equipment have some part or parts that conduct electricity and other parts that do not. A bare overhead distribution line is held up by insulators and also utilizes the air around it for insulation. Transformer windings have insulation around each turn of the conductors and, in some cases, use oil, in addition, to raise the insulation value between the conductor and the grounded components.

The primary factor that determines the level of insulation that is required is the operating voltage. Other factors, such as current and frequency, also play a part; however, they are secondary to voltage. Therefore, the first consideration in testing insulation is whether it can support the required voltage without breakdown. This is accomplished by measuring the leakage current that flows through the insulation medium when a voltage is applied.

There are almost as many types of insulation as there are different applications. There are several things, however, that they all have in common. Moisture and contamination decrease an insulator's ability to withstand voltage and increase the amount of leakage current that will flow. The insulation will also deteriorate with age. Overheating causes deterioration to be greatly accelerated. A common rule of thumb is that the life expectancy of the insulation is cut in half for every 10 °C above its rating that the equipment operates.

Another characteristic that some types of insulation have is that, as the voltage rises, the insulation will maintain its integrity until it reaches the point of failure. Then, near the point of failure, the insulating capability drops very rapidly, often accompanied by an arc or puncture in the insulation.

6.2.1 DC tests

The most common method of testing insulation integrity is to apply a dc voltage and measure the leakage current. The insulation resistance is then determined by dividing the voltage by the current. There are many commercially available test instruments that have specific voltage outputs and provide the readings directly in ohms. This is referred to as "insulation resistance testing," and frequently as "Megger testing" or "Meggering."

For low-voltage equipment, common test voltages are 100 V, 250 V, 500 V, and 1000 V. Test instruments are also available that have test voltages of 2500 V, 5000 V, and 10 000 V for use on medium-voltage equipment. For most testing beyond 10 000 V, the test equipment no longer has a fixed output voltage. This is considered "high-potential" testing, commonly called "hi-pot" testing; and the voltage is continuously adjustable so that it can be ramped up slowly. The leakage current is usually measured directly when hi-pot testing is being performed, since the voltage is no longer fixed and getting a direct readout in ohms would be more difficult.

Basic electrical theory helps explain how dc insulation resistance testing works and what information it provides. A capacitor can be defined as a device for storing electrical charge that consists of two conductors separated by an insulator. From this definition, it can be seen that many electrical components, such as cables, bus bars, motor windings, and transformer windings, have capacitance to ground, since the conductors of these items are separated from ground by an insulating medium. The amount of capacitance is determined by the sizes of the conductors, their proximity to each other, and the electrical properties of the insulation. When dc voltage is applied between the conductor and ground, the capacitor that is formed by the bus, cable, etc., and ground charges up. Once this charging is completed, the remaining dc current (called "leakage current") that is flowing is due to the dc resistance of the insulation. By dividing the voltage by the leakage current, the value of the insulation resistance is obtained. It is common practice to measure the insulation resistance over a specific interval (to allow the capacitance to charge) and then record the final reading.

6.2.1.1 Insulation resistance tests

Insulation resistance tests are typically performed on motors, circuit breakers, transformers, low-voltage (unshielded) cables, switchboards, and panel boards to determine if degradation due to aging, environmental, or other factors has affected the integrity of the insulation. This test is normally conducted for 1 min, and the insulation resistance value is then recorded.

As mentioned earlier, the electrical properties of the insulation and the amount of surface area directly affect the capacitance between the conductor and ground, and therefore affect the charging time. With larger motors, generators, and transformers, a common test is to measure the "dielectric absorption ratio" or the "polarization index" of the piece of equipment being tested. The dielectric absorption ratio is the 1 min insulation resistance reading divided by the 30 s insulation resistance reading. The polarization index is the 10 min (continuous) insulation resistance reading divided by the 1 min reading.

Both of these provide additional information as to the quality of the insulation. Many types of insulation become dry and brittle as they age, thereby becoming less effective capacitors. Thus, a low polarization index (less than 2.0) may indicate poor insulation. Even though insulation may have a high insulation resistance reading, there could still be a problem, since the motor and transformer windings are subjected to strong mechanical stresses on starting. With the exception of electronic equipment (which can be damaged by testing), insulation resistance testing is normally done on most types of new equipment and is also part of a maintenance program. It is a good practice to perform insulation resistance testing on switchgear and panelboards after maintenance has been performed on them, just prior to re-energizing them. This prevents re-energizing the equipment with safety grounds still applied or with tools accidentally left inside.

6.2.1.2 High-potential testing

High-potential testing, as its name implies, utilizes higher levels of voltage in performing the tests. It is generally utilized on medium-voltage (1000–69 000 V) and on high-voltage (above 69 000 V) equipment (see Figure 6-1). As stated earlier, the leakage current is usually measured. In some cases, such as in cable hi-potting, the value of leakage current is significant and can be used analytically. In other applications, such as switchgear hi-potting, it is a pass/fail type of test, in which sustaining the voltage level for the appropriate time (usually 1 min) is considered "passing."

6.2.1.3 Medium- and high-voltage cable testing

Most cables that are rated for use at voltage levels above 600 V are shielded cables. A shielded cable has a conductor in the center, a semiconducting layer over the strands that is

Source: electro • test inc.

Figure 6-1—Circuit breaker high-potential testing

surrounded by insulation, a semiconducting layer, and then a metal foil or wire mesh that surrounds the whole assembly. There is usually another layer over the shield that makes up the outer jacket of the cable. It is a common practice to hi-pot test the cables on initial installation in order to verify that the cables were not damaged when they were pulled into place and that all the splices and/or terminations were installed properly. The voltage level that is selected usually is lower than factory test levels, frequently 80% of the dc equivalent of the factory test level.

There are normally two considerations that are given to hi-pot testing of cables as a routine maintenance practice. One is a function of the chosen maintenance philosophy [i.e., breakdown maintenance, preventive maintenance, predictive maintenance, or reliability-centered maintenance (RCM)]. The other depends upon the type of operation and how critical it is to have continuous power without interruption.

The debate on whether or not to perform maintenance hi-pot testing centers around the fact that a cable in marginal condition can be caused to fail by the hi-pot test itself. A cable that is in good condition should not be harmed. People who subscribe to maintenance testing feel that it is much better to have the cable fail under test. Cable maintenance testing frequently is performed at 50–65% of the factory test voltage. Problems can then be corrected while the circuit is intentionally shut down, thus avoiding an in-service failure that could interrupt production. It is important to remember that the necessary material, such as splice kits or cable terminations, should be available to facilitate repairs should the cable fail during testing.

6.2.2 AC tests

The most common ac insulation test is ac hi-pot testing at 60 Hz. The capacitance of the insulation is a large factor in ac testing. With dc insulation testing, the capacitance of the insulation charges up over time and the residual leakage current is an indication of the resistance of the insulation. This is not true of ac insulation testing. Since the voltage is changing at 60 Hz, the leakage current may be predominantly the capacitive charging current of the insulation under test. For this reason, ac hi-pot testing is usually a pass/fail type test in which "passing" means that the insulation was capable of holding the required test voltage, usually for 1 min. A common rule of thumb in the U.S. for determining the value of ac test voltage for switchgear and panelboards is to apply twice the rated voltage plus 1000 V. This comes from UL 891-1994 [B55].[1] The practice in some other countries is to limit routine maintenance testing to 80% of this value in order to prevent the overstressing of the insulation through repetitive testing.

Typically, switchgear, panelboards, and bus ducts are tested using ac voltage. AC hi-pot testing has the advantage that it stresses the voids and air gaps more than dc, and thus provides a better test on switchgear and panelboard insulation. Because of the high capacitance of large motors, transformers, and long shielded cables, however, ac hi-potting may not be practical due to the limitations of the power capability of ac test equipment. The ac hi-pot has to have the capability of continuously providing the required charging and leakage current at the test potential. DC is normally used in place of ac for applications with a large capacitance.

6.2.3 Power-factor testing dielectric loss angle (DLA) testing

Power-factor testing is a special type of ac insulation testing. In power-factor testing, the phase relationship between the applied test voltage and the resulting leakage current is determined.

A special test set is used that supplies voltage and current at 60 Hz (see Figure 6-2). The volt-amperes and watts are measured with the test set and the power factor is determined.

The primary use of power-factor testing has been on medium- and high-voltage transformers, circuit breakers, and bushings, particularly when they are oil filled. Because of the high dielectric strength of oil, a dc insulation resistance test may not detect a problem unless the equipment is about to fail. Power-factor testing provides a means to evaluate the condition of the whole or parts of the insulation system and to track the deterioration of the insulating system as it ages. Many firms have contributed test data on transformers, bushings, and circuit breakers for several decades, thus creating historic trends on many of the common makes of equipment. This supplies additional data for evaluating a piece of equipment that has been in service and for which no prior test data exists. This test provides an extremely useful tool to help predict equipment failure so that it can be scheduled for replacement prior to failure.

---

[1]The numbers in brackets correspond to those of the bibliography in 6.8.

Utilities across the U.S. have been power-factor testing their large transformers and circuit breakers for many years and trending the information to help determine replacement or major repair needs.

Source: electro • test inc.

Figure 6-2—Insulation power-factor testing

6.3 Protective device testing

A "protective device" is generally referred to as any piece of equipment whose primary function is to detect an electrical abnormality or failure, such as an overload or fault, and to automatically initiate the removal of the problem. A fuse, for example, senses excessive current flow and "blows" to interrupt the flow of current. Some relays would be classified as "protective" while other relays would be classified as "control" or "alarm," depending on what action the relay initiates. A relay would be considered a protective relay if it trips open a circuit breaker, switch, or motor contactor when an electrical fault occurs. A relay that operates a light or an annunciator, on the other hand, would not normally be considered a protective relay, but would be considered an alarm relay.

The most common types of protective devices are protective relays, fuses, and circuit breakers with integral trip units. Fuses and most low-voltage circuit breakers are self-contained, where the sensing and operating devices are all in a common housing. Medium- and high-voltage circuit breakers, motor starters, and spring-loaded switches have separate protective devices, such as relays, that work together as a system to perform the overall protective function. Typically, protective relays sense voltage and/or current and operate a circuit breaker, motor starter, or spring-loaded switch when a predetermined level of voltage, current, power, frequency, etc., is reached.

6.3.1 Protective relay testing

Protective relaying is a very broad subject. Only a brief overview can be given here.

There are two major objectives in protective relaying. First, a protective relay serves to provide equipment protection (i.e., locate and isolate overloads, short circuits, undervoltages, and other types of electrical problems quickly in order to minimize damage). Second, the protective device that is closest to the problem should operate first to clear the problem, and no other device should operate unless the closest one fails. This concept, known as "selective tripping" or "selectivity," maintains service to as much of the electrical system as possible by isolating only the problem area. In order to achieve these objectives, each relay must function as it was designed, and the relays must function in conjunction with the other protective devices in the system. Having all the protective devices function as one overall protective system is called "coordination."

Each protective device has specific parameters within which it has been designed to operate. For example, a single element fuse has a value of current above which it opens. It takes a specific amount of time for a given current to melt the link away and open the fuse. Manufacturers of fuses publish "time-current" curves that show how long it takes a fuse to operate for varying current values. Generally, the higher the current, the shorter the time.

This same inverse current-vs.-time concept is used for overcurrent relays and for low-voltage circuit breakers. Relays and low-voltage circuit breakers (with internal trip units) have a range of "pickup" operating current that causes them to operate. In many cases, this value of current is adjustable. By properly selecting the type, characteristic, and/or setting of fuses, relays, or circuit breakers, the system can be coordinated so that the device that is closest to the problem opens before any device upstream of it. It is necessary to select compatible time-current characteristics of the devices for the entire system, in addition to selecting the proper settings for the devices.

Prior to performing protective relay testing, a coordination study should be completed to determine the proper settings for the relays to be calibrated. This is usually done by the design engineer when the system is first installed. If there have been revisions or additions to the system, a new study may be necessary.

Once the coordination study has been completed, the relays need to be calibrated to the proper settings. There are special test sets available for this purpose that inject currents and voltages, as necessary, and time the various operations of the relays. This type of testing is usually performed by a technician who specializes in this area. Depending upon the relay to be calibrated, quite complex test equipment may be required and in-depth training in protective relaying may be needed to properly set the relay.

6.3.2 Low-voltage circuit breakers

Low-voltage circuit breakers come in the following three major types:

a) Power (air-frame) circuit breakers;
b) Molded-case circuit breakers;
c) Insulated-case circuit breakers.

Power circuit breakers start with a frame size of 600 A and go up to 4000 A. The sensing unit that operates the breaker on a short circuit or overload may be either an oil-dash pot with springs and copper coils (for older breakers) or may consist of current transformers (CT) and an electronic trip unit. With the advent of the electronic trip unit, the number of possible settings and trip functions has dramatically increased, making it easier to coordinate circuit breakers with other protective devices.

Molded-case circuit breakers and insulated-case circuit breakers are very similar in mechanical construction and insulation. The circuit breakers' contacts and operating mechanisms are totally enclosed in a molded plastic housing. The difference between the two is that a molded-case circuit breaker normally has a thermal-magnetic trip unit (i.e., a trip unit made up of two pieces: a thermal unit to sense overload that uses two dissimilar metals and a magnetic unit to trip on short circuit), while an insulated-case circuit breaker has CTs and an electronic trip unit built into the insulated case.

The most thorough test for all three types of circuit breakers is by "primary injection." A special test set that puts out high (fault level) current at low voltage (typically 6–20 V ac) is used to functionally test the circuit breaker. These test sets have built-in timing functions; therefore, the breaker can be tested at various currents in order to make sure that it operates within the time-current specifications that are provided by the manufacturer and that it is calibrated to perform in conformance with the coordination study.

For circuit breakers that have electronic trip units, it is often possible to do "secondary injection" testing. This is usually done with a special test set that is designed for the trip unit. It injects low-level test currents into the trip unit, directly testing only the trip unit. For this reason, primary injection testing is a better practice, as it tests the whole circuit breaker (CTs, shunt trip, etc.) in a manner that is similar to how the breaker would operate during a fault.

In addition to testing the tripping characteristics of the circuit breaker by injecting current, it is also normal practice to test the insulation resistance (usually at 1000 V dc) and the resistance of the breaker's contacts. The contact resistance can be measured directly with a low-resistance ohmmeter (usually in milliohms) or indirectly by performing a millivolt drop test. A millivolt drop test is performed by using a primary injection test set to inject rated continuous current through the breaker while measuring the millivolt drop across the breaker's poles. It is a comparative test between each phase of the breaker in which the millivolt reading typically should not differ by more than 50% between phases.

Power circuit breakers have mechanical adjustments and inspections that should also be periodically checked. The manufacturer's instructions list the adjustments for each model.

6.3.3 Instrument transformers

There are two common designations of instrument transformers: CTs and voltage transformers (VTs) or potential transformers (PTs). The function of an instrument transformer is to reduce the level of voltage or current so that the protective relay (or metering) does not have to be rated for full line voltage or current.

The insulation resistance, transformer ratio, and polarity may be tested in both CTs and VTs. The ratio is the number of turns of wire in the primary winding divided by the number of turns of wire in the secondary winding. The polarity is determined by which way the wire was wrapped around the iron core. This determines the relationship between the primary winding terminal (H1) and the secondary winding terminal (X1) so that X1 is positive with respect to X2 at the same time that H1 is positive with respect to H2. The correctness of polarity is important to the correct operation of many relays and metering instruments.

CTs often have two additional tests performed: "burden" and "saturation" tests. The burden on a CT is the amount of impedance connected to the secondary winding as a load, usually in the form of protective relays or metering. The burden test consists of injecting a known current level (usually 1–5 A ac) into the load (usually from the shorting terminal block of the CT) and measuring the voltage at the point of injection. The impedance (or burden) of the circuit is the ratio of the voltage measured to the current injected.

A saturation test is performed to find out the voltage at which the iron in the CT saturates. A known voltage source is connected to the secondary of the transformer and is raised in steps, while the current value is recorded at each step. When saturation is reached, the given voltage changes cause much smaller changes in current.

The saturation test is used in conjunction with the burden test to make sure that the CT is capable of operating the load (usually protective relays) to which it may be subjected. If the burden on the CT is too high, it may go into saturation and be unable to maintain its proper ratio. When this happens, protective relays may trip too slowly or not at all due to an insufficient level of current from the CT secondary.

6.4 Analytical tests

Analytical tests address specific aspects of a piece of equipment or system under test. The tests are usually more specialized to the equipment under evaluation and may utilize test equipment that is designed for that specific test.

6.4.1 Winding and contact resistance

Winding and contact resistance are similar in that both are looking for a very low ohmic value, since they are measuring the "resistance" of a component that is supposed to conduct electricity. A Kelvin Bridge has long been a standard method of measuring low values of resistance, and is still in use today. With the advent of electronics, there are digital meters available that also are capable of measuring very low values (milliohms or microohms) of resistance. The typical low-resistance ohmmeter uses four terminals (to eliminate lead resistance) in which a dc current is injected into the conductor to be measured and the voltage drop across the conductor is measured.

Contact resistance test sets can be used to measure the resistance of bus joints and cable joints, as well as the closed contacts of a circuit breaker or motor starter (see Figure 6-3). In many cases, it is a comparative type test in which the resistance of one set of contacts is compared to the readings obtained from the other two phases of the same, or a similar, piece of equipment.

Source: electro • test inc.

Figure 6-3—Contact resistance testing

Winding resistance differs from contact resistance in that the inductance of large windings can interfere with the operation of the test set. There are test sets, available commercially, that are designed specifically for large transformer and motor windings, for cases in which a standard low-resistance ohmmeter is not adequate.

6.4.2 Transformer turns ratio (TTR) testing

The voltage across the primary of a transformer is directly proportional to the voltage across the secondary, multiplied by the ratio of primary winding turns to secondary winding turns.

In order to ensure that the transformer was wound properly when it was new, and to help locate subsequent turn-to-turn faults in the winding, it is common practice to perform a TTR test. The simplest method would be to energize one primary winding with a known voltage (that is less than or equal to the winding's rating) and measure the voltage on the other winding. Since source test voltages can fluctuate, it is often more accurate to use a test set, designed for this purpose, that creates the test voltage internally, thus giving a direct read-out of the ratio measured.

6.4.3 Motor surge comparison testing

Motor surge comparison testing addresses the problem of insufficient test voltage to find the weak insulation between turns by utilizing a high-voltage pulse. Two identical high-voltage pulses are introduced into two windings of a motor. The propagation of the pulse through one winding is compared to the propagation of the identical pulse through the winding next to it. An oscilloscope (usually built into the surge tester) is used to look at the traces and to compare them. The patterns should be identical (or very nearly so) and can appear as one trace (two superimposed traces) if both windings are good. A turn-to-turn failure (or a failure to ground) is indicated by two distinctly different traces appearing on the oscilloscope.

Motor surge comparison testing has been used by motor winding shops for many years. There are now portable models that are available for field testing. Motor surge comparison testing has proven to be a valuable tool in detecting the early stages of a winding failure, both from the standpoint of preventing an unexpected failure during operation and preventing a catastrophic failure of the motor so it can be repaired instead of needing to be replaced.

6.4.4 Time-travel analysis

Time-travel analysis is primarily performed on medium- and high-voltage circuit breakers. The test consists of plotting the voltage that is applied to the trip coil (or close coil for closing tests) and the movement of the breaker mechanism vs. time with a high-speed recorder such as a light beam oscillograph. The plot is then analyzed to determine opening (or closing) speed and contact bounce, which may be critical in the proper operation of the breaker. These tests can also be used to determine if the contacts of each phase opened (or closed) simultaneously.

6.4.5 Infrared scanning

Infrared scanning is a method that is utilized to locate high-resistance connections ("hot spots") by using a camera that turns infrared radiation into a visible image (see Figure 6-4). This test is performed with the equipment in service carrying normal load current, which is a major advantage because it does not interrupt normal production. Exposure to energized equipment, of course, carries the possibility of exposure to electrical hazards. The operator shall recognize and deal with such potential hazards accordingly.

The most common use of infrared scanning is to locate loose or corroded connections in switchboards, panel boards, bus ways, and motor starters. It is a comparative type test in which the person who performs the scan is looking for an area that appears brighter (hotter) than a similar area, such as a lug connection on phase "A" as it compares to similar connections on phases "B" and "C." The person should be aware of how unbalanced loading may affect heating, thereby giving an indication similar to looseness.

One limitation of infrared scanning is that the equipment has to be carrying enough load for the hot spots to be visible. At lower loads, there may not be enough heat generated to locate a problem, even when the connections are significantly looser than they should be.

Source: electro • test inc.

Figure 6-4—Infrared inspection

6.4.6 Oil testing

Many medium- and high-voltage transformers and circuit breakers utilize different types of oils for insulation. Chemical testing of the oil has proven to be a very dependable method of locating existing or potential problems. Only a brief overview of some of the common tests can be provided here.

One of the most obvious problems that would significantly reduce the insulation value of the oil is contamination such as moisture or, for circuit breakers and load top changers, carbon. This can be tested on-site by measuring the voltage at which dielectric breakdown occurs with a special test set that is designed for this purpose.

Oil samples may be sent to a testing laboratory for a series of tests. Measurements of the acidity give an indication of how much oxidation or contamination the oil has experienced. Interfacial tension, the force that is required to rupture the surface tension at an oil-water interface, is also an indication of possible oxidation or contamination.

One of the most successful tests in determining if a transformer winding has experienced hot spots, corona discharge, or arcing is the dissolved gas analysis test. An oil sample is taken with a special cylinder that is air tight, and the gases that are dissolved in the oil are analyzed. By determining the type and amount of gas that has been dissolved in the oil, predictions can be made about the internal integrity of the transformer winding.

6.5 Grounding tests

Grounding is a broad subject that has an entire standard dedicated to it. (See IEEE Std 142-1991 [B29].)

A facility grounding system can be divided into the following three major subsystems:

a) Power system grounding for fault and personnel protection, as required by the National Electric Code® (NEC®) (NFPA 70-1996) [B53];
b) Signal reference grounding;
c) Lightning protection.

Each of these subsystems has a different function to accomplish, but they all should be compatible. In most cases there can be multiple interconnections between the subsystems, and care should be exercised to prevent one subsystem from disrupting the other.

Power system grounding is for the purpose of minimizing the electrical hazard that the power distribution system poses to people. The NEC [B53] has very specific grounding requirements to accomplish this objective, and IEEE Std 142-1991 [B29] provides recommended practices. As it is used here, the term "power system grounding" would include "system grounding" and "equipment grounding" from IEEE Std 142-1991 [B29].

Signal reference grounding refers to the use of ground as a reference for electronic controls and communication. In many instances, electronic equipment uses its own metal frame or case as the signal reference. The NEC [B53] requires the outside of metal enclosures of electrical equipment to be bonded to power system ground (by the use of "equipment grounding conductors") to prevent shock hazards. Therefore signal reference is connected to power system ground. The objective becomes the design of a grounding system so that the signal reference does not pick up stray ground currents from the distribution system, thereby causing operating problems for the electronic equipment.

Lightning protection is intended primarily to dissipate the energy from a lightning strike in a manner that is safe for personnel and that causes the least amount of equipment damage. The lightning protection system may have multiple interconnections with building steel and the power system ground. Since lightning is a cloud to earth phenomenon, the resistance between the lightning protection/power system ground point and the outside earth is an important factor.

From the above discussion, it should be noted that two separate factors relate to grounding. The first is how well the lightning protection/power system ground is connected to the earth, and the second is how well and in what manner the equipment grounding conductors and other metal surfaces connect the electrical equipment throughout the facility to the signal reference/lightning protection/power system ground. (Contrary to some equipment manufacturers' literature, the resistance between signal reference and the earth is rarely an important issue.)

When do we care how much resistance there is between the grounding system and the earth? When is it important for this value to be low? The answers to these two questions become obvious from the next question, "When is the earth conducting part of the electrical current?" This can happen when lightning strikes a building. The earth may also be part of the electrical circuit during certain line-to-ground faults if the equipment grounding conductor is faulty or missing, or if the earth provides a low-impedance path back to the source of power.

When the earth is not part of the conducting path, the issue becomes the amount of impedance between the individual pieces of equipment that are connected by the grounding system. This can be an important issue for both safety-related grounding and signal-reference grounding. Higher impedance means greater potential difference.

One reason that the NEC [B53] requires 208 Y/120 V and 480 Y/277 V wye systems to be solidly grounded and equipment grounding conductors to be installed is to prevent the outside of electrical equipment from becoming energized and creating a potential shock hazard. If the grounding system has a high resistance, a failure to ground of the insulation inside of the equipment may not allow enough current to flow to operate the protective device (fuse or circuit breaker) and de-energize the faulted equipment. In such a situation, a person coming in contact with the outside case could be exposed to a shock hazard. A correctly sized and installed equipment grounding conductor provides a low impedance for the fault current so that the protective device operates properly, thereby isolating the faulted equipment.

With signal reference grounding, the issue is usually one of electromagnetic interference. A low-impedance grounding system limits the potential that is developed between neutral and ground, which can cause operational problems for sensitive electronic equipment. It also provides a separate path for stray ground currents to flow on, thus preventing interference that could disrupt the communication or control signals between electronic equipment operating at high frequency and low magnitude.

6.5.1 Grounding electrode test

The NEC [B53] uses the term "grounding electrode" for the electrical conductor or ground rod that is buried in the earth. The test utilized to determine the resistance between a grounding electrode and the earth is called a "fall-of-potential" test. It is performed by connecting a test set that provides a known current between the grounding electrode to be tested and an auxiliary electrode driven in the ground for the purpose of testing (see Figure 6-5). The voltage is then measured between the electrode under test and a second auxiliary electrode. This "voltage" second auxiliary electrode is driven into the ground between the other two in a sampling of places. The resistance that is obtained from these measurements should have a flat point in the curve when plotted vs. the distance between the test rod and the current rod. The resistance to earth is determined by dividing the voltage by the current.

Source: electro • test inc.

Figure 6-5—Ground resistance testing

6.5.2 Two-point resistance tests

Two-point resistance tests are used to measure the resistance of the equipment grounding conductor and its bonding of the electrical equipment. This type of test is very similar to the low-resistance measurements that are taken for contact resistance tests, and the same test equipment can be used. The purpose of this test is to measure the resistance of a test specimen to a known reference point.

6.5.3 Soil resistivity test

Soil resistivity is a measure of resistance per unit length of a uniform cross-section of earth, usually expressed in ohm-centimeters. It is performed with a four-terminal test set that uses four equally spaced electrodes, driven into the ground. The current flows between the outside two, and the voltage is measured between the inside two. The earth resistivity is equal to the reading of the test set, times 2Π, times the distance between two stakes (in centimeters).

6.6 Functional testing

Functional testing consists of simulating various normal and abnormal conditions, and monitoring the system performance for proper operation. This can be as simple as opening and closing a circuit breaker electrically, or as complex as performing transient stability tests on an emergency generator. Functional testing is slightly different when done from a maintenance viewpoint as compared to acceptance testing.

If proper acceptance testing is performed on the initial start-up of the system, installation problems such as the miswiring of alarm or control circuits should be found and easily corrected.

The problems a maintenance engineer may encounter usually pertain to deteriorated equipment, adjustment problems, modified designs, and system expansions that have been added since the initial installation. The protective devices should be operated to ensure that they function properly. Where emergency generators and/or transfer switches are part of the electrical distribution, there is no better functional test of the entire system than tripping the normal source circuit breaker and observing the system for proper operation.

If the electrical distribution system was not acceptance tested when initially installed, much more extensive functional testing should be performed during initial maintenance periods. Common problems that are frequently encountered include neutral conductors that are grounded downstream of the main bonding jumper (on low-voltage ground-fault systems), improper CT polarity, problems in the control and alarm circuits, and wiring errors between switchboard sections where wiring was revised or completed in the field.

6.7 Testing procedures and specifications

There are many sources of information on testing and maintenance of electrical equipment. The magnitude alone can be a formidable barrier to establishing an effective maintenance program, since it would literally take years to sort through all the applicable references and apply the material to a specific set of circumstances. To avoid this and assist the reader to quickly get a handle on particular needs, an overview of most applicable references is provided in this chapter, along with a detailed bibliography.

6.7.1 Sources of electrical equipment testing and maintenance information

The following is an alphabetical listing of the groups that provide procedures and specifications for electrical testing and maintenance:

a) American National Standards Institute (ANSI);
b) American Society for Testing and Materials (ASTM);
c) Association of Edison Illuminating Companies (AEIC);
d) Institute of Electrical and Electronics Engineers (IEEE);
e) Insulated Cable Engineers Association (ICEA);
f) InterNational Electrical Testing Association (NETA);
g) National Electrical Manufacturers Association (NEMA);
h) National Fire Protection Association (NFPA);
i) Occupational Safety and Health Administration (OSHA).

6.7.2 NFPA 70B-1998, Electrical Equipment Maintenance [B54]

The development of NFPA 70B began in 1968 with the Board of Directors of the National Fire Protection Association who established a committee "to develop suitable texts relating to preventive maintenance of electrical systems and equipment used in industrial-type applications with the view of reducing loss of life and property. The purpose is to correlate generally applicable procedures for preventive maintenance that have broad application to the more common classes of industrial electric systems and equipment without duplicating or superseding instructions that manufacturers normally provide."

This document provides reasoning for performing preventive maintenance, economic justification discussions, and detailed procedures for maintenance activities. Maintenance procedures are given for typical items of electrical equipment.

NFPA 70B-1998 [B54] provides the core of an excellent preventive maintenance program by providing a great deal of information for direct use by the facility engineers and maintenance personnel. With the addition of specific instructions that are provided in the manufacturer's literature, the vast majority of electrical equipment maintenance needs are covered.

6.8 Bibliography

Additional information may be found in the following sources:

[B1] Accredited Standards Committee C2-1997, National Electrical Safety Code® (NESC®).[2]

[B2] AEIC CS5-87-1994, Cross-linked Polyethylene Insulated Shielded Power Cables Rated 5 Through 46 kV.[3]

[B3] AEIC CS6-87-1996, Ethylene Propylene Rubber Insulated Shielded Cables Rated 5 Through 69 kV.

[B4] ANSI C37.20-1969, American National Standard for Switchgear Assemblies—Including Metal-Enclosed Bus (withdrawn).[4]

[B5] ASTM D877-87 (1995), Standard Test Method for Dielectric Breakdown Voltage of Insulating Liquids Using Disk Electrodes.[5]

[B6] ASTM D923-97, Standard Practices for Sampling Electrical Insulating Liquids.

---

[2]The NESC is available from the Institute of Electrical and Electronics Engineers, 445 Hoes Lane, P.O. Box 1331, Piscataway, NJ 08855-1331, USA (http://www.standards.ieee.org/).

[3]AEIC publications are available from the Association of Edison Illuminating Companies, 600 N. 18th Street, P. O. Box 2641, Birmingham, AL 35291-0992, USA (http://www.aeic.org/). AEIC publications are also available from Global Engineering Documents, 15 Inverness Way East, Englewood, Colorado 80112-5704, USA (http://global.ihs.com/).

[4]ANSI C37.20-1969 has been withdrawn; however, copies can be obtained from Global Engineering, 15 Inverness Way East, Englewood, CO 80112-5704, USA, tel. (303) 792-2181 (http://gobal.ihs.com/).

[5]ASTM publications are available from the American Society for Testing and Materials, 100 Barr Harbor Drive, West Conshohocken, PA 19428-2959, USA (http://www.astm.org/).

[B7] ASTM D924-92, Standard Test Method for Dissipation Factor (or Power Factor) and Relative Permittivity (Dielectric Constant) of Electrical Insulating Liquids.

[B8] ASTM D971-91, Standard Test Method for Interfacial Tension of Oil Against Water by the Ring Method.

[B9] ASTM D974-97, Standard Test Method for Acid and Base Number by Color-Indicator Titration.

[B10] ASTM D1298-85 (1990) e1, Standard Practice for Density, Relative Density (Specific Gravity), or API Gravity of Crude Petroleum and Liquid Petroleum Products by Hydrometer Method.

[B11] ASTM D1500-96, Standard Test Method for ASTM Color of Petroleum Products (ASTM Color Scale).

[B12] ASTM D1524-94, Standard Test Method for Visual Examination of Used Electrical Insulating Oils of Petroleum Origin in the Field.

[B13] ASTM D1533-96, Standard Test Methods for Water in Insulating Liquids (Karl Fischer Reaction Method).

[B14] ASTM D1816-97, Standard Test Method for Dielectric Breakdown Voltage of Insulating Oils of Petroleum Origin Using VDE Electrodes.

[B15] ASTM D2129-97, Standard Test Method for Color of Clear Electrical Insulating Liquids (Platinum-Cobalt Scale).

[B16] ASTM D2285-97, Standard Test Method for Interfacial Tension of Electrical Insulating Oils of Petroleum Origin Against Water by the Drop-Weight Method.

[B17] ASTM D3284-90a, Standard Test Method for Combustible Gases in the Gas Space of Electrical Apparatus in the Field.

[B18] ASTM D3612-96, Standard Test Method for Analysis of Gases Dissolved in Electrical Insulating Oil by Gas Chromatography.

[B19] Dakin, T. W., "Electrical insulation deterioration treated as a chemical rate phenomenon," *AIEE Transactions,* vol. 67, pp. 113–122, Apr. 1948.

[B20] Gill, A. S., *Electrical Equipment Testing & Maintenance.* Englewood Cliffs, NJ: Prentice Hall.

[B21] IEEE Standards Collection: Power and Energy—Switchgear and Substations Set, 1998 Edition.[6]

[B22] IEEE Distribution, Power, and Regulating Transformers Standards Collection, 1998 Edition.

[B23] IEEE Surge Protection Standards Collection, 1995 Edition.

[B24] IEEE Std 43-1974 (Reaff 1991), IEEE Recommended Practice for Testing Insulation Resistance of Rotating Machinery.

[B25] IEEE Std 48-1996, IEEE Standard Test Procedures and Requirements for Alternating-Current Cable Terminations 2.5 kV Through 765 kV.

[B26] IEEE Std 81-1983, IEEE Guide for Measuring Earth Resistivity, Ground Impedance, and Earth Surface Potentials of a Ground System (Part 1).

[B27] IEEE Std 95-1977 (Reaff 1991), IEEE Recommended Practice for Insulation Testing of Large AC Rotating Machinery with High Direct Voltage.

[B28] IEEE Std 141-1993, IEEE Recommended Practice for Electric Power Distribution for Industrial Plants (*IEEE Red Book*).

[B29] IEEE Std 142-1991, IEEE Recommended Practice for Grounding of Industrial and Commercial Power Systems (*IEEE Green Book*).

[B30] IEEE Std 242-1986 (Reaff 1991), IEEE Recommended Practice for Protection and Coordination of Industrial and Commercial Power Systems (*IEEE Buff Book*).

[B31] IEEE Std 399-1997, IEEE Recommended Practice for Industrial and Commercial Power Systems Analysis (*IEEE Brown Book*).

[B32] IEEE Std 400-1991, IEEE Guide for Making High-Direct-Voltage Tests on Power Cable Systems in the Field.

[B33] IEEE Std 421.3-1997, IEEE Standard for High-Potential Test Requirements for Excitation Systems for Synchronous Machines.

[B34] IEEE Std 446-1995, IEEE Recommended Practice for Emergency and Standby Power Systems for Industrial and Commercial Applications (*IEEE Orange Book*).

[B35] IEEE Std 450-1995, IEEE Recommended Practice for Maintenance, Testing, and Replacement of Vented Lead-Acid Batteries for Stationary Applications.

[B36] IEEE Std 493-1997, IEEE Recommended Practice for the Design of Reliable Industrial and Commercial Power Systems (*IEEE Gold Book*).

[6]IEEE publications are available from the Institute of Electrical and Electronics Engineers, 445 Hoes Lane, P.O. Box 1331, Piscataway, NJ 08855-1331, USA (http://www.standards.ieee.org/).

[B37] IEEE Std 602-1996, IEEE Recommended Practice for Electric Systems in Health Care Facilities (*IEEE White Book*).

[B38] IEEE Std 637-1985 (Reaff 1992), IEEE Guide for the Reclamation of Insulating Oil and Criteria for Its Use.

[B39] IEEE Std 739-1995, IEEE Recommended Practice for Energy Management in Industrial and Commercial Facilities (*IEEE Bronze Book*).

[B40] IEEE Std 1106-1995, IEEE Recommended Practice for Maintenance, Testing, and Replacement of Nickel-Cadmium Storage Batteries for Generating Stations and Substations.

[B41] IEEE Std C37.13-1990 (Reaff 1995), IEEE Standard for Low-Voltage AC Power Circuit Breakers Used in Enclosures.

[B42] NEMA AB 1-1993, Molded-Case Circuit Breakers.[7]

[B43] NEMA AB 2-1984, Procedures for Field Inspection and Performance Verification of Molded-Case Circuit Breakers Used in Commercial and Industrial Applications.

[B44] NEMA AB 4-1996, Guidelines for Inspection and Preventive Maintenance of Molded-Case Circuit Breakers Used in Commercial and Industrial Applications.

[B45] NEMA MG 1-1993, Motors and Generators.

[B46] NEMA SG 3-1995, Low Voltage Power Circuit Breakers—Power Switching Equipment.

[B47] NEMA WC 2-1980, Steel Armor and Associated Coverings for Impregnated-Paper-Insulated Cables.

[B48] NEMA WC 3-1992, Rubber-Insulated Wire and Cable for the Transmission and Distribution of Electrical Energy.

[B49] NEMA WC 4-1988, Varnished-Cloth-Insulated Wire and Cable for the Transmission and Distribution of Electrical Energy.

[B50] NEMA WC 5-1992, Thermoplastic-Insulated Wire and Cable for the Transmission and Distribution of Electrical Energy.

[B51] NEMA WC 7-1988, Cross-Linked-Thermosetting-Polyethylene-Insulated Wire and Cable for the Transmission and Distribution of Electrical Energy.

[7]NEMA publications are available from Global Engineering Documents, 15 Inverness Way East, Englewood, Colorado 80112, USA (http://global.ihs.com/).

[B52] NEMA WC 8-1988, Ethylene-Propylene-Rubber-Insulated Wire and Cable for the Transmission and Distribution of Electrical Energy.

[B53] NFPA 70-1996, National Electrical Code® (NEC®).[8]

[B54] NFPA 70B-1998, Recommended Practice for Electrical Equipment Maintenance.[9]

[B55] UL 891-1994, Dead-Front Switchboards (DoD).[10]

[B56] Westinghouse Electric Corporation, *Maintenance Hints,* WEC, Pittsburgh, Pa.

[8]The NEC is available from Publications Sales, National Fire Protection Association, 1 Batterymarch Park, P.O. Box 9101, Quincy, MA 02269-9101, USA (http://www.nfpa.org/). Copies are also available from the Institute of Electrical and Electronics Engineers, 445 Hoes Lane, P.O. Box 1331, Piscataway, NJ 08855-1331, USA (http://www.standards.ieee.org/).

[9]NFPA publications are available from the National Fire Protection Association, Batterymarch Park, Quincy, MA 02269, USA (http://www.nfpa.org/). Copies are also available from the Sales Department, American National Standards Institute, 11 West 42nd Street, 13th Floor, New York, NY 10036, USA (http://www.ansi.org/).

[10]UL standards are available from Global Engineering Documents, 15 Inverness Way East, Englewood, Colorado 80112, USA (http://global.ihs.com/).

Chapter 7
Introduction to electrical safety

7.1 General discussion

NFPA 70E-1995[1] defines an electrical hazard as "a dangerous condition, such that, inadvertent or unintentional contact, or equipment failure, can result in shock, arc flashburn, thermal burn, or blast." Electrical safety is defined in NFPA 70E-1995 as "recognizing hazards associated with the use of electrical energy and taking precautions so that hazards do not cause injury or death."

Some explanation is necessary to fully appreciate the words of the two definitions. Personnel should understand the nature and consequences of electrical hazards and the reasons for practicing electrical safety. The nature of electrical hazards is discussed in 7.2. The consequences are demonstrated by some true case histories of injuries, deaths, and near-misses, as described in 7.3. Finally, 7.4 discusses the reasons for spending money and effort to be electrically safe.

It is important to understand the four main phases of protection from electrical hazards.

a) First, electrical installations should be designed and constructed to be safe by complying with the criteria of recognized and generally accepted good engineering practices. This subject is discussed in Chapter 9.

b) Second, the integrity of electrical equipment shall be maintained with particular emphasis on enclosures, insulation, operating mechanisms, grounding, and circuit protective devices. Maintenance is discussed in Chapters 5 and 6, as well as in Chapter 9.

c) Third, unless there are serious overriding circumstances, electrical equipment shall be placed in an electrically safe work condition before personnel work on or near it. Safe practices shall be used to establish an electrically safe work condition. An electrically safe work condition is discussed in Chapter 10.

d) Fourth, safe work practices and adequate protective equipment, tools, and test equipment shall be understood and used when it is not feasible to establish an electrically safe work condition, or when de-energizing would create a greater hazard of another kind. Safe practices are covered in Chapter 10. Personal protective equipment, appropriate tools, and other protective methods are discussed in Chapter 11. The safe use of electrical equipment is discussed in Chapter 12.

The details of the above concepts should be studied and documented into a cohesive electrical safety program. Chapter 8 discusses establishing such a program.

These electrical safety chapters (Chapters 7 through 12) give guidance as to what needs to be done to avoid creating dangerous conditions in the first place, how to recognize them when they do exist, and how to take appropriate precautions to avoid having anyone injured or killed.

---

[1] Information on references can be found in 7.6.

7.1.1 Acronyms and abbreviations

CPR	cardiopulmonary resuscitation
D-N	Drouet-Nadeau
NEC	National Electrical Code®
NEMA	National Equipment Manufacturers Association
NESC	National Electrical Safety Code®
NFPA	National Fire Protection Association
OSHA	Occupational Safety and Health Administration
UL	Underwriters' Laboratories

7.2 Exposure to electrical hazards

The definition of an electrical hazard indicates that hazards can result from poor physical condition of equipment or facilities, sometimes simply called "unsafe conditions." It also indicates that injuries can result from the careless or inadvertent actions of people, sometimes simply called "unsafe acts." Taking precautions means that safety considerations regarding such conditions and actions should be given to all aspects of electrical work, starting from the initial design concept, through installation and start-up, and continuing on into the post-operational maintenance activities.

There are essentially three recognized kinds of injury that may result while working on or near electrical hazards:

a) Electrical shock;
b) Burns from contact, arcs, or flashes;
c) Impact from blasts.

These injuries are described in detail in 7.2.1 through 7.2.3.

7.2.1 Electrical shock

The electrical hazard of which most people are aware, and the one that most electrical safety standards have been built around, is the electrical shock hazard. Electrical shock affects human beings in the following ways:

a) Currents as small as a few milliamperes through the heart can cause disruption of the natural electrical signals that the heart uses to perform its normal functions. Voltage levels as low as 50 V with low skin resistance and current flowing through the chest area can cause fibrillation, which can result in death. This is why all personnel who routinely work on or near energized electrical parts should be familiar with cardiopulmonary resuscitation (CPR). They may be required to perform CPR on a fellow worker who has experienced an electrical shock.
b) Electrical shock can damage human tissue where the current enters and exits the body. Within the body, the current can also damage internal body parts in its path. The degree of damage is dependent on the amount of current, the type of contact, the duration of contact, and the path of the current through the body.
c) Electrical shock causes the muscles to contract. Due to muscle contraction, the person experiencing the shock may not be able to release the conductor causing the shock (known as the "let-go threshold"). This grasping leads to longer exposure.

Several standards offer guidance regarding safe approach distances in order to minimize the possibility of shock from exposed electrical conductors of different voltage levels. The most recent, and probably the most authoritative, guidance is presented in NFPA 70E-1995. The limits of approach to exposed energized electrical conductors are discussed in Chapter 10 of this guide.

Most electrical personnel are aware that there is a danger of electrocution. Few really understand, however, just how small a quantity of electric energy is required for electrocution. Actually, the current drawn by a 7.5 W, 120 V lamp, passed from hand to hand or foot, is enough to cause fatal electrocution. Just as it is current, and not voltage, that heats a wire, it is current that causes physiological damage.

Different values of 60 Hz ac and their effects on a 68 kg (150 lb) human are listed in Table 7-1. In short, any current of 10 mA or more may be fatal. Those between 75 mA and 4 A can be fatal from heart disruption. Those above 5 A may be fatal from severe internal or external burns. It is a fact, however, that shocks in this last current range are statistically less likely to be fatal than those in the 75 mA to 4 A range. In view of the wide diversity of injuries derived from contact with electrical energy, it is only logical that, to prevent electrical shock or electrocution, exposure to energized parts should be minimal, if required at all.

Table 7-1—Current range and effect on a 68 kg (150 lb) human

Current (60 Hz)	Physiological phenomena	Feeling or lethal incidents
<1 mA	None	Imperceptible.
1 mA	Perception threshold	
1–3 mA		Mild sensation.
3–10 mA		Painful sensation.
10 mA	Paralysis threshold of arms	Cannot release hand grip. If no grip, victim may be thrown clear. (May progress to higher current and be fatal.)
30 mA	Respiratory paralysis	Stoppage of breathing (frequently fatal).
75 mA	Fibrillation threshold, 0.5% (greater than or equal to 5 s exposure)	Heart action discoordinated (probably fatal).
250 mA	Fibrillation threshold, 99.5% (greater than or equal to 5 s exposure)	Heart action discoordinated (probably fatal).
4 A	Heart paralysis threshold (no fibrillation)	Heart stops for duration of current passage. For short shocks, heart may restart on interruption of current (usually not fatal from heart dysfunction).
>5 A	Tissue burning	Not fatal unless vital organs are burned.

Source: Lee [B14].

As mentioned previously, the effects of shock are dependent upon the amount and path of current that flows through the body. That current, in turn, is related to the applied voltage and the resistance of the path of the current. Table 7-2 shows some typical values of skin-contact resistance under various conditions. Table 7-3 shows typical resistance values of various materials. Figure 7-1 provides a ready means of evaluating the physiological effect of a human resistance and a 60 Hz voltage source. Note that at about 600 V, the resistance of the skin ceases to exist; it is simply punctured by the high voltage just like capacitor insulation. For higher voltages, only the internal body resistance impedes the current flow. It is usually somewhere around 2400 V that burning becomes the major effect. Below this voltage, fibrillation and/or asphyxiation are the usual manifestations.

Table 7-2—Human resistance values for various skin-contact conditions

Condition	Resistance (in Ω)	
	Dry	Wet
Finger touch	40 000–1 000 000	4000–15 000
Hand holding wire	15 000–50 000	3000–6000
Finger-thumb grasp[a]	10 000–30 000	2000–5000
Hand holding pliers	5000–10 000	1000–3000
Palm touch	3000–8000	1000–2000
Hand around 3.8 cm (1.5 in) pipe (or drill handle)	1000–3000	500–1500
Two hands around 3.8 cm (1.5 in) pipe	500–1500	250–750
Hand immersed	—	200–500
Foot immersed	—	100–300
Human body, internal, excluding skin	200–1000	

[a] Data interpolated.

Source: Lee [B14].

Table 7-3—Resistance values for 130 cm^2 areas of various materials

Material	Resistance (in Ω)
Rubber gloves or soles	>20 000 000
Dry concrete above grade	1 000 000–5 000 000
Dry concrete on grade	200 000–1 000 000
Leather sole, dry, including foot	100 000–500 000
Leather sole, damp, including foot	5000–20 000
Wet concrete on grade	1000–5000

Source: Lee [B14].

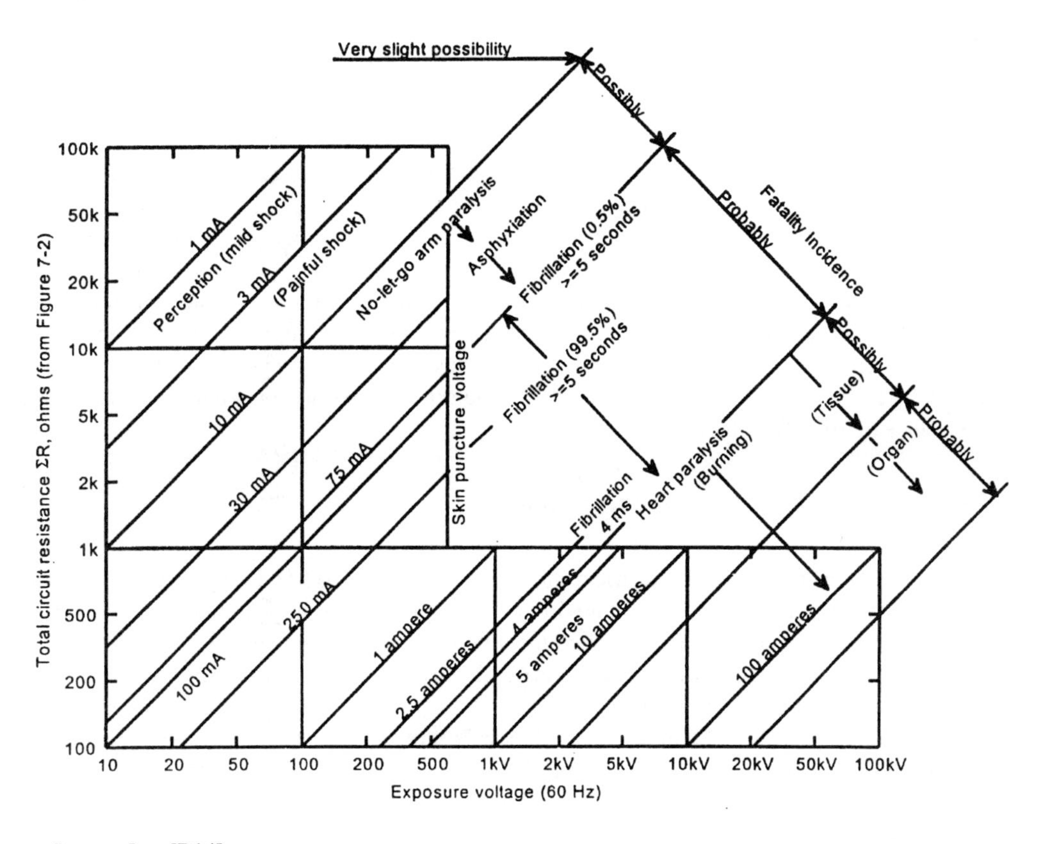

Source: Lee [B14].

Figure 7-1—Resistance—voltage—current effect appraisal chart

In fibrillation, the victim may not recover consciousness. On the other hand, the victim may be conscious, deny needing help, walk a few feet, and then collapse. Death may occur within a few minutes, or may take hours. Detection of the fibrillation condition requires medical skill. The application of closed-chest massage, a treatment in which blood is circulated mechanically in a fibrillation victim, can result in the death of a subject whose heart is not in fibrillation.

In Figure 7-1, the fibrillation line is shown at 75 mA. Actually, this is the threshold sensitivity for an exposure of 5 s or more. For shorter times, the threshold current is higher, along a constant I^2t line. It should be noted that, if the duration of shock is only 0.004 s, or one-quarter cycle of 60 Hz, the fibrillation threshold current is raised to over 2500 mA or 2.5 A. Referring again to Figure 7-1, such a situation moves the threshold over almost to the heart-paralysis and tissue-burning zone.

This sensitivity, increasing with time, explains why a victim who is "frozen" to a current source is much more likely to be electrocuted than one whose contact does not involve hand grasp. A full hand grasp immobilizes the victim such that he cannot let go; hence, the exposure time may extend to many seconds, placing it in range of the 75 mA threshold. In comparison, a casual contact (such as with a fingertip) causes instant retraction of the arm, thereby interrupting the shock current path. In this case, the victim is exposed for only a few thousandths of a second, and is much less likely to sustain an injury.

In addition, data has been compiled that shows the corresponding voltage required to force certain current values through a person who has a circuit resistance of 500 Ω. Although this value appears to be rather low for human body resistance, it can be approached by someone who has sweat-soaked cloth gloves on both hands and a full hand grasp of a large energized conductor and a grounded pipe or conduit. Moreover, cuts, abrasions, or blisters on hands can negate the skin resistance, leaving only internal body resistance to oppose current flow. A circuit value as low as 37.5 V could be dangerous in this instance. For a person with a resistance of 1000 Ω, 75 V could be dangerous; 120 V would be dangerous for a human body with a resistance of 1600 Ω or less.

7.2.2 Burns from contact, arcs, or flashes

Almost everyone is aware that electrical shock can be a hazard to life. Many people, however, have experienced minor shocks with no dire consequences. This tends to make people somewhat complacent around electricity. What most people don't know is that approximately half of the serious electrical injuries involve burns. Electrical burns include not only burns from contact, but also radiation burns from the fierce fire of electric arcs that result from short circuits due to poor electrical contact or insulation failure. The electric arc between metals is, next to the laser, the hottest thing on earth. It is about four times as hot as the sun's surface. Where high arc currents are involved, burns from such arcs can be fatal, even when the victim is some distance from the arc. Serious or fatal burns can occur at distances of more than 304 cm (10 ft) from the source of a flash. In addition to burns from the flash itself, clothing is often ignited. Fatal burns can result because the clothing cannot be removed or extinguished quickly enough to prevent serious burns over much of the body.

Thus, even at what a person thinks to be a large distance, serious or fatal injuries can occur to a person's bare skin or skin covered with flammable clothing as a result of a severe power arc. Electrical workers are frequently in the vicinity of energized parts. It is only the relative infrequency of such arcs that has limited the number of injuries. Examples of exposure are working on open panelboards or switchboards, hook stick operation of medium-voltage fuses, testing of cable terminals, grounding before testing, or working in manholes near still-energized cables.

Several studies, tests, and technical papers are being written on the subject of the flash hazard. Safety standards and procedures are being developed to recognize the fact that arcs can cause serious injuries at significant distances from energized sources. Equally important in these new safety standards is the fact that, in many cases, only trained people with arc protective equipment should approach exposed energized electrical equipment. Spectators should stay away because, even though they think they are far enough away, they generally don't have an understanding of what is a safe approach distance. (Refer to Table 2-3.3.5 of NFPA 70E-1995 for safe approach distances.) Depending upon the fault energy available, spectators can be seriously hurt at large distances from the point of an arc.

7.2.2.1 Nature of arcs

Electrical arcing is the term that is applied to the passage of substantial electrical currents through what had previously been air. It is initiated by flashover or the introduction of some conductive material. Current passage is through ionized air and the vapor of the arc terminal material, which is usually a conductive metal or carbon. In contrast to current flow through low-pressure gases such as neon, arcing involves high temperatures of up to, or beyond, 20 000 °K (35 000 °F) at the arc terminals. No materials on earth can withstand these temperatures; all materials are not only melted, but vaporized. Actually, 20 000 °K (35 000 °F) is about four times as hot as the surface temperature of the sun.

The vapor of the terminal material has substantially higher resistance than solid metal, to the extent that the voltage drop in the arc ranges from 29.53 V/cm (75 V/in) to 39.37 V/cm (100 V/in), which is several thousand times the voltage drop in a solid conductor. Since the inductance of the arc path is not appreciably different from that of a solid conductor of the same length, the arc current path is substantially resistive in nature, thus yielding unity power factor. Voltage drop in a faulted large solid or stranded conductor is about 0.016–0.033 V/cm (0.5–1 V/ft).

For low-voltage circuits, an arc length of 29.53–39.37 V/cm (75–100 V/in) consumes a substantial portion of the available voltage, leaving only the difference between source voltage and arc voltage to force the fault current through the total system impedance, including that of the arc. This is the reason for the "stabilization" of arc current on 480 Y/277 V circuits when the arc length is of the order of 10.16 cm (4 in), such as with bus spacing.

For higher voltages, the arc lengths can be substantially greater, e.g., 2.54 cm (1 in) per 100 V of supply, before the system impedance starts to regulate or limit the fault current. Note that the arc voltage drop and the source voltage drop add in quadrature, the former resistive, the latter substantially reactive. The length or size of arcs in the higher voltage systems thus can be greater and can readily bridge the gap from energized parts to ground or other polarities with little drop in fault current.

The hazard of the arc is not only due to the level of voltage. Under some cases it is possible to generate a higher energy arc from a lower voltage than from a higher voltage. The amount of arc energy generated is dependent upon the amount of short-circuit current available and the amount of time before the fault causing the arc is cleared (removed from the power source) by a circuit breaker or fuse.

7.2.2.2 The arc as a heat source

The electric arc is widely recognized as a very high-level source of heat. Common uses are arc welding, electric arc furnaces, and even electric cauterizing of wounds to seal against infection while deeper parts are healing. The temperatures at the metal terminals are extraordinarily high, being reliably reported at 20 000 °K (about 35 000 °F). One investigator reports temperatures as high as 34 000 °K (about 61 000 °F), and special types of arcs can reach 50 000 °K (about 90 000 °F). The only higher temperature source known on earth is the laser, which can produce 100 000 °K (about 180 000 °F).

The intermediate (plasma) part of the arc, i.e., the portion away from the terminals, is reported as having a temperature of 13 000 °K (about 23 500 °F). In comparison, the surface temperature of the sun is about 5000 °K (9000 °F), so the terminal and plasma portions are 4 and 2.5 times, respectively, as hot as the sun's surface. The temperature below the surface of the sun is, of course, much higher; it is approximated at 10 000 000 °K (18 000 000 °F) at the center.

7.2.2.3 Development of arc size

In a bolted fault there is no arc, so little heat will be generated in that case. Should there be appreciable resistance at the fault point, temperature could rise to the melting and boiling points of the metal, and an arc could be started. The longer the arc becomes, the more of the available system voltage it consumes. Consequently, less voltage is available to overcome the supply impedance, and the total current decreases.

7.2.2.4 Effect of temperature on human tissue and clothing

The human body can exist in only a relatively narrow temperature range that is close to normal blood temperature, which is around 36.5 °C (97.7 °F). Survival much below this level requires insulation with clothing. Temperatures that are slightly above this level can be compensated for by perspiration. Studies show that at skin temperature as low as 44 °C (110 °F), the body temperature equilibrium mechanism begins to break down in about 6 h. Cell damage can occur beyond 6 h at that temperature. Between 44 °C (110 °F) and 51 °C (124 °F), the rate of cell destruction doubles for each 1 °C (1.8 °F) temperature rise. Above 51 °C (124 °F), the rate is extremely rapid. At 70 °C (158 °F), only a 1 s duration is sufficient to cause total cell destruction.

7.2.3 Impact from blasts

The rapid expansion of air caused by a fault current has been recognized for some time as one of the electrical hazards. What was not recognized previously was the fact that this blast and its effects could be calculated, and precautions against its effects could be taken.

As in shock and arc damage, the principal thing to do to avoid this hazard is to stay away from exposed energized electrical systems. Unfortunately, electrical work is a profession that sometimes requires exposure to this hazard. The total exposure can be reduced, however, by not using switchgear rooms for offices, warehouses, or smoke-break rooms. Sometimes, electrical equipment must be maintained and operated while it is energized. The risk of a fault occurring while people are in close proximity to this equipment shall be taken into account.

7.2.3.1 Pressures developed by arc blasts

People are exposed to two dangers from electrical arcs: burns and blasts. The blast can cause falls and other injuries, as well as damage nearby structures. A relationship is developed between arc current and pressure for an applicable range of distance.

For familiarization with some units used for pressures in the SI (metric) system, the following may be useful:

— 1 newton (N) = 0.2248 pounds force (lbf)
— 1 newton/m^2 = 0.0209 lb/ft^2
— 1 atmosphere (atm) = 2 116 lb/ft^2 = 1.0125×10^5 N/m^2

7.2.3.2 Arc forces defined

Reports of the consequences of electrical power arcs in air include descriptions of the rearward propulsion of personnel who were close to the arc. In many cases, the affected people do not remember being propelled away from the arc, often failing to remember the arc occurrence itself. The relative infrequency of power arcs has tended to distract interest from determining the nature and magnitude of this pressure.

Also, the heat and molten metal droplet emanation from the arc can cause serious burns to nearby personnel, a fact that has also tended to reduce interest in the rearward propulsion and pressures generated.

Another consequence of arcs is structural damage. One power arc in a substation of the Quebec Hydroelectric System caused collapse of a nearby substation wall. To determine the magnitude of pressure that is generated by the arcing fault, M. G. Drouet and F. Nadeau of the Institut de Recherche de l'Hydro-Québec were assigned to develop theoretical and practical bases for this phenomenon. The results of their work are described in a 1979 paper (see Drouet and Nadeau [B5]).

Drouet and Nadeau's work shows that actual pressures are an order of magnitude greater than theoretical values. According to a reviewer, Dr. Nettleson, this phenomenon is attributed to a very high frequency component of pressure that is not recorded by measuring apparatus. Regardless of this, the measured amplitudes of pressures from a 100 kA, 10 kV arc reached about 2×10^4 N/m^2 (400 lb/ft^2) at a distance of 1 m (3.3 ft). This pressure is about ten times the value of wind resistance that walls are normally built to withstand, so such an arc could readily destroy a conventional wall at distances of about 12 m (40 ft) or less. A 25 kA arc could similarly destroy a wall at a distance of 3 m (9.5 ft).

Pressures on projected areas of individuals at 0.6 m (2 ft) from a 25 kA arc would be about 7750 N/m^2 (160 lb/ft^2). This is sufficient to place a force on the front of a person's body of about 2100 N (480 lbf). Such pressures are also found to be damaging to human ears. Mandatory hearing protection, as in other high noise-level locations, should be considered during the hazard analysis (see 10.3.5 and 10.3.6.).

7.2.3.3 Development of arc pressure

The pressures from an arc are developed from two sources: the expansion of the metal when boiling and the heating of the air by passage of the arc through it. Copper expands by a factor of 67 000 in vaporizing, similar to the way that water expands about 1670 times when it becomes steam. This accounts for the expulsion of near-vaporized droplets of molten metal from an arc. These are propelled for distances of up to about 3 m (10 ft). It also generates plasma (ionized vapor) outward from the arc for distances proportional to the arc power. With copper, 53 kWs vaporizes 0.328 cm^3 (0.05 in^3), thereby producing 54 907 cm^3 (3350 in^3) of vapor. Therefore, 16.39 cm^3 (1 in^3) of copper vaporizes into 1.098 m^3 (38.8 ft^3) of vapor.

The air in the arc stream expands in warming up from its ambient temperature to that of the arc, or about 20 000 °K (35 000 °F). This heating of the air is related to the generation of thunder by the passage of lightning current through it. Dr. R. D. Hill developed theoretical pressures at distances of 0.75–4 cm (0.295–1.575 in) from a 30 kA peak lightning strike. These pressures ranged from 40 atm down to 9 atm. Dr. Hill's data are plotted in Figure 7-2, on log-log scale, and the straight-line of these points is extrapolated to 100 cm (39.37 in) distance, at which distance the pressure would have been 0.45 atm. Multiply this 0.45 by 200/30, to match the peak power of the Drouet-Nadeau (D-N) tests, and the Hill data becomes 3.3 atm, rather close to the D-N theoretical value of 2.7 atm.

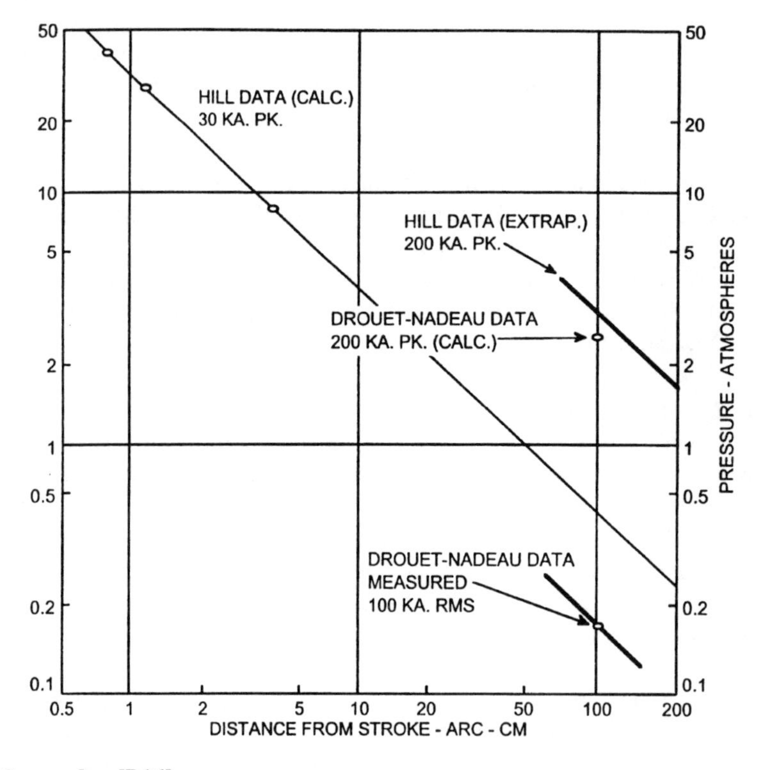

Source: Lee [B16].

Figure 7-2—Pressure vs. distance from stroke or arc

The actual measured pressure by D-N from a 200 kA peak, 100 kA rms current was 0.19 atm or 0.07 times the calculated theoretical pressure. Since this is the only available measured pressure level, it is used to generate a family of lines, shown as Figure 7-3. In this figure, pressures are shown for arc currents from 1 kA to 100 kA rms, for a range of distances of 15 cm to 30 m (0.5 –100 ft), from the arc center to the point of interest. From this, the pressure may be determined for a 25 kA arc at a distance of 60 cm (2 ft) to be 7656 N/m^2 (160 lb/ft^2). This pressure has at least one useful aspect—personnel close to an arc are propelled rapidly away from the heat source, substantially reducing the degree to which they are subjected to burns.

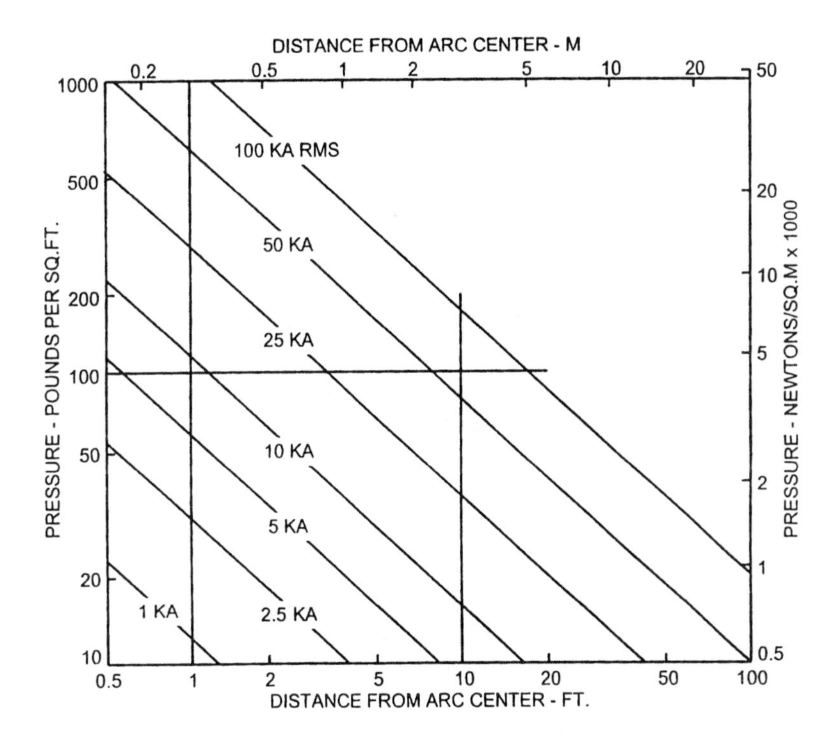

Source: Lee [B15].

Figure 7-3—Pressure vs. distance from arc center

The hot vapor from the arc starts to cool immediately. While hot, however, it combines with the oxygen of the air, thus becoming the oxide of the metal of the arc. These continue to cool, solidify, and become minute particles in the air, appearing as black smoke for copper and iron, and gray smoke for aluminum. They are still quite hot, and cling to any surface they touch, actually melting into many insulating surfaces they may contact. Many people think that these are carbon particles. The oxide particles are very difficult to remove because surface rubbing is not effective. Abrasive cleaning is necessary on plastic insulation. A new surface varnish should be applied, or surface current leakage could occur and cause failure within days.

Persons exposed to severe pressure from proximity to an arc are likely to suffer short-time loss of memory and may not remember the intense explosion of the arc itself. This is a brief concussion that interferes with the transfer from short-time memory to long-time memory. This phenomenon has been found true even for high-level electrical shocks.

It is evident that persons working in conditions where power arcing is possible should be protected not only against arc burns, but also against falling and ear damage.

7.3 Case histories

As the following incidents show, a variety of things can go wrong when working on or near electrical conductors and circuit parts, even when they are thought to be de-energized. These incidents demonstrate how important it is to strictly control such work by using only trained and qualified personnel who know and use safe practices and appropriate protective equipment. The incidents in 7.3.2 prove that the use of proper protective equipment does save lives and protect employees from injury.

7.3.1 Incidents resulting in injury

7.3.1.1 Case no. 1—shock

A mechanic, while working on some equipment in the rear of a power-type circuit breaker auxiliary metering compartment, accidentally came in contact with adjacent energized transformer terminals. He apparently had not checked for other energized components in the vicinity in which he was working. He required medical treatment for shock.

7.3.1.2 Case no. 2—shock

A construction electrician was assigned to a job installing wiring for lighting fixtures in a 208 Y/120 V system. The job was already partially completed. The electrician was told by the foreman that everything was "dead." The "home run" was already installed, and those wires were protruding from a junction box. The electrician was on a ladder and began to remove the insulation from the home run wires using "wire skinners" with one hand. His other hand was holding the wires, but was also in contact with the box. He was not wearing gloves, and received a severe shock causing him to fall from the ladder. Fortunately, he received only minor bruises from the fall. He later said, "I should have checked myself to see that it was really dead."

7.3.1.3 Case no. 3—severe shock

A field engineer was using portable radiography equipment to inspect the quality of medium-voltage cable terminations. A construction electrician was assisting him. The radiography equipment consisted of a cathode ray tube in a metallic casing (called an X-ray head) and a control unit. A control cable interconnected these units. The engineer was on top of a stepladder, leaning against the switchgear enclosure, grasping the handles of the X-ray head, and adjusting its position. The electrician proceeded to plug the control unit into a 120 V outlet and connect the control cable. Suddenly, the engineer received a severe shock and fell off of the ladder. It was later determined that the plug-in connection between the control cable and the control unit was poorly polarized. The electrician had forced the connection together in the wrong orientation, putting 120 V on the casing of the X-ray head. The engineer was taken to the medical department at the site, and was observed for the remainder of the day. He said his grasp was locked onto the handles momentarily until his weight broke him loose during his fall. He complained of muscular problems for several days afterward.

7.3.1.4 Case no. 4—shock and burn

An electrician was part of a crew that was operating a 13 kV switch in a new electric power system. One switch blade did not close properly. As other crew members packed up their tools and prepared to return to the shop to discuss the problem with supervision, they heard a noise. They turned around to see the switch door open, and the electrician surrounded by smoke and flames. While trying to read a part number, the electrician had accidentally made contact with an energized part of the switch, and a flash had occurred. A coworker pulled the man from the switch and revived him using cardiopulmonary resuscitation. Nine months later, after having some fingers and toes amputated, the electrician was still receiving medical treatment as a result of the accident.

7.3.1.5 Case no. 5—shock and burn

A construction electrician had been running wiring all day long for several new heating and air conditioning units in the space above a suspended ceiling. The electrical feeder that supplied the new installation was a 480 Y/277 V system, but it had been verified as de-energized earlier in the day; however, no lockout had been installed. The electrician, at one point late in the day, began to prepare the main feeder for connection. He was working on a ladder and started to remove the insulation from one of the wires. He did not have gloves on. He let out a yell as he became "hung up" on the wires. Apparently, sometime during the day, someone had turned on the feeder breaker. A carpenter, working right beside the electrician, recognized that the electrician was "frozen" to the wires due to muscle contraction. The carpenter used a piece of lumber to knock the electrician off the ladder, breaking him free of the wire. The electrician fell onto the floor, rolled, and banged against a wall. He was taken to the hospital and kept for two days for treatment and observation, having received a severe shock, second degree burns to both hands, and facial cuts from the fall.

7.3.1.6 Case no. 6—flash, blast, and burns

A contract electrician had finished drying out a 480 V bus duct. While working on a ladder, reinserting the plug-in units onto the bus, he encountered difficulty getting one of the units to make up properly. He banged on the plug-in unit and was met with a flash and blast that severely burned him and knocked him off the ladder. He was not wearing any protective clothing.

7.3.1.7 Case no. 7—fatal shock

An electrician was installing a 277 V lighting fixture in an industrial plant. Turning the power off was inconvenient, as it involved winding his way around and through some vessels and pipes, and going up two flights of stairs to the lighting panel. Apparently thinking that there was not much danger in this "minor" job, he didn't bother to turn the power off. Working in an existing junction box, he made up two of the connections, and was trying to remove an existing wire nut to install the third wire. He was having difficulty removing the nut, so he used a lug crimping tool to try to pull and twist the wire nut off. He gripped the tool too hard and it cut through the outer shell of the wire nut, making contact with the energized wire inside. He was electrocuted.

7.3.1.8 Case no. 8—fatal shock and blast

A contract electrical maintenance crew had arrived the night before a planned shutdown of some low- and medium-voltage equipment. The plant engineer offered to show the foreman the equipment on which they would be working the next day. He opened the doors to the medium-voltage equipment, leaving the doors open as he went down the line. Two members of the crew trailed the plant engineer and the foreman, and were intrigued with some discoloration on a cable terminal wrapping in the first cubicle. One electrician, not recognizing that the cubicle was bottom fed, assumed that the cable was de-energized, and approached it too closely, possibly even touching it. He was electrocuted and, at the same time, initiated a blast that severely impacted the second electrician. The second electrician was later unable to remember what happened.

7.3.1.9 Case no. 9—burns

Power switchboard operators and a shift foreman were racking in a 2.4 kV circuit breaker. As they were trying to raise the breaker into position with the dc elevator motor, they experienced trouble and burned up the elevator motor in the process. The shift foreman, who had limited experience with this particular switchgear, called to get some assistance racking in the gear. Another foreman and operator arrived. Since trouble had been experienced racking the breakers in and out before, the operators thought that possibly the breaker was either rusty from lack of regular operation or affected by climate problems. They decided to continue racking the breaker in manually. As the breaker got to within 5.08 cm (2 in) of being fully racked in, a flash from the cubicle occurred. Fortunately, the operator racking in the breaker had a full flash suit on, and his burns were minimized. The foreman who was a good distance from the cubicle, but had no flash suit on when it flashed, received a slight burn on one arm and his face as he moved to help the operator. This incident would have been a fatality if the operator had not worn a flash suit. The switchgear behind the operator had the Bakelite nameplate burned off. The foreman who was back from the switchgear and was not suited up was injured more than the operator in the flash suit.

7.3.1.10 Case no. 10—burns

An electrician heard a hissing sound coming from within some 480 V switchgear. He opened the door to the switchgear to get a better view. He did not have on a flash suit, only side-shield glasses and a hard hat. The gear flashed phase-to-phase right in front of him. He almost died as a result of the flash. He spent months in the hospital recovering from severe burns.

7.3.1.11 Case no. 11—burns

Three people were burned when the 2.4 kV motor contactor they were racking in tried to close when the stabs made contact with the bus. The unit was set on auto-start position and as soon as the control power transformer became energized, the contactor picked up, causing arcing at the stabs. The resultant ionized gas then caused a phase-to-phase fault. The workers were not wearing flash suits.

7.3.1.12 Case no. 12—burns

A set of dc switches was not operated in the proper sequence. This resulted in a direct short when one of the switches was closed. An arc occurred across the switch as the operator was closing it. The operator received burns to his neck and arms, even though he was wearing required flash protection. The injury was due in part to the way that the operator had put on his flash suit. He did not properly close the hood and jacket.

7.3.1.13 Case no. 13—fatal burns

An electrician was spray-washing a 480 V switch with a solvent. The garden sprayer nozzle made contact across the open switch. There was a large amount of fault current available at this point in the circuit. The electrician was fatally burned when an arc flashed. The temperature may have reached 20 000 °K (35 000 °F). A wooden handrail that was 3.05 m (10 ft) from the switch ignited.

7.3.1.14 Case no. 14—burns and blast

A shopping center was being enlarged and new switch units were being added to a 480 V metal-clad substation power center. A new vertical bus was added to an existing vertical section to accommodate new switch units. The bolts attaching the vertical bus to the riser from the main horizontal bus were installed so that the threaded ends were toward the back of the switch unit. Not only that, but the bolts were about 2.54 cm (1 in) longer than required to secure the bus to the riser. The net result was that the bolt ends were only about 0.16 cm (1/16 in) back of the new switch enclosure. An electrician and a helper were inserting a 400 A, 600 V fuse in a new fusible switch. In pressing the fuse into the fuse mounting, the box was deflected back into short-time contact with the bolt ends. This initiated a power fault of about 100 kA rms. The electrician was seriously burned by the infrared radiation from the arc, and his clothes were ignited by the molten copper droplets expelled from the arc. He was propelled backwards by approximately 2670 N (600 lbf) of pressure on his chest from the fault, falling against the front of another section of switchgear that was 2.74 m (9 ft) from the one he had been working on. The helper, who had been watching him at the open door of the switch enclosure, was propelled backward about 7.5 m (25 ft), completing two backward somersaults before ending up against a wall. He was not injured, and hurried back to the electrician, helping him up and extinguishing the flames of his clothes. A calculation indicated that the electrician had been propelled backwards nearly 0.6 m (2 ft) in 0.1 s, which substantially reduced the radiation burning.

7.3.1.15 Case no. 15—blast and flash burn

Electricians were installing new wiring in an older plant. One horizontal plug-in bus duct was to be fed through an existing empty conduit from an existing air-break switch. The conduit from the switch to the bus duct was attached to the top of the switch enclosure, and the line terminals of the switch were energized. The electricians were passing a fish tape from the bus duct end toward the switch, but they could not push it the entire distance. One of the electricians then took another fish tape and started to push it up from inside the switch to hook it onto the hook end of the first fish tape. He did not de-energize the line terminals of the switch or put any guarding material over the energized terminals. In pushing on the second tape, it buckled toward, and touched, the energized terminal, thus initiating a ground fault arc from an energized terminal to the fish tape, which was grounded to the inside of the conduit. The electrician was knocked down backwards, and extensively burned.

7.3.1.16 Case no. 16—blast

Two men in a plant were walking away from a newly closed 480 V switch, when it exploded with a heavy arc. Though about 2.4 m (8 ft) away from the switch, one man received substantial arc droplet spot melting on his nylon jacket. Both men had their hearing adversely affected. One continued to experience pain and required medical care for over 14 months after the incident.

7.3.1.17 Case no. 17—blast

In an area of the country known for a large amount of snowfall, a number of outdoor substations had been experiencing 480 V bus faults, particularly when springtime came. Investigation showed severe water drippage onto the interior components, due to the condensation of warmer air against the underside of the still-cold top of the switchgear. The substation heaters were not operative due to heater burnout. Where the water droplets struck bus and breaker insulation, tracking was initiated, starting arcs. One of these faults occurred when a plant electrician was standing about 0.9 m (3 ft) from the front of the 480 V switchgear. The arc fault was on the buses, behind the breaker fronts. Pressure through the opening below the bottom breakers, however, propelled the electrician back against the substation fence, 2.1 m (7 ft) from the switchgear. Fortunately, he was not seriously injured.

7.3.1.18 Case no. 18—electrocution

A fast food restaurant worker was kneeling on the floor to insert the male plug of a portable electric toaster into a 120 V, 20 A receptacle that was mounted in a grounded metal enclosure with a cover. The floor was wet because it had been recently mopped. As the worker held the cover open with his left hand, and began to insert the male plug into the receptacle outlet, his index finger apparently touched the energized prong of the plug. The victim was found convulsing, and he died shortly afterwards.

7.3.1.19 Case no. 19—double electrocutions

Two painters were using an aluminum extension ladder to paint a metal light pole. One worker was standing on the ladder painting, and his coworker was on the ground holding the ladder. The ladder slipped away from the pole and contacted a 12 460 V overhead power line which was very near the pole. Both painters were electrocuted.

7.3.1.20 Case no. 20—double electrocutions and burns

Three workers were using a telescoping boom crane to move a section of steel framing at a construction site. As the section was moved, it came in contact with a 23 000 V overhead power line. Two of the three workers who were in direct contact with the load were electrocuted while the third received serious electrical burns.

7.3.2 Incidents where protection prevented injury

7.3.2.1 Case no. 21—flash, no injury

An operator was in the process of closing a disconnect switch on a 480 V, Size 4, NEMA 1 starter when it faulted and flashed. The operator, who was wearing the required flash protection, had stood to one side, and had turned away while operating the switchgear. The operator was not injured, but his protective gloves were scorched by molten copper. The switchgear door was blown open, and the door and switchgear on the opposite side of the aisle were burned.

7.3.2.2 Case no. 22—flash and blast, no injury

An operator was sent out to clear and tag Pump A for maintenance. He shut Pump A down, blocked the valves, and then went to the motor control center to open and tag the switch. He mistakenly opened the switch to Pump B, which was running loaded. Arcing in the switch caused a flash and blast, thus blowing the switch compartment door open. Fortunately, the operator was wearing a flash suit and was not hurt.

7.3.2.3 Case no. 23—flash and blast, no injury

A fuse mechanism on a 2300 V non-load-break fused disconnect faulted phase-to-phase when closed in with a hot stick. The employee was wearing a flash suit. The hood face shield and jacket had molten copper spots. No injury resulted from the fault.

7.3.2.4 Case no. 24—flash, no injury

An operator attempted to close a pole-mounted disconnect switch, feeding a 14 kV oil circuit breaker, with the circuit breaker in the closed position. The breaker had been left in the closed position following the testing of a bushing that had been replaced. An operator and a supervisor had visually inspected the breaker before operating the disconnect switch and thought the breaker was open. The breaker position indicator signs were over-sprayed with paint and the breaker was mistakenly assumed to be open when it was closed. The operator put on a full flash suit with rubber gloves and proceeded to close the disconnect. He was standing under the 14 kV disconnect using the gang operator to close the disconnect. As the disconnect was closed, an electrical arc occurred. The resultant fault broke the insulator, and molten copper and porcelain came down on the operator. A protective relay system cleared the arc. No injury occurred as a result of this incident.

7.3.2.5 Case no. 25—flash and blast, no injury

An operator was in the process of energizing a motor from a push-button start switch on some 480 V switchgear. When he flipped the start switch, which was located on the door of the starter unit, the switchgear arced and blew up. He was wearing protective equipment and was not hurt. He immediately put out the resulting fire. The switchgear and its associated light and wiring were checked for faults. No conclusive evidence of the cause was found, but sandblasting work had been done earlier in the area of the switchgear.

7.4 Reasons for practicing electrical safety

There are at least three good reasons for practicing electrical safety:

a) Personal reasons, which affect everyone as caring individuals and employers;
b) Business reasons, because safety makes good business sense;
c) Regulatory and legal reasons, because violations can result in fines and/or imprisonment.

7.4.1 Personal considerations

The first reason for practicing electrical safety is the adverse way in which we are personally affected by electrical accidents, especially since we know that they are often avoidable. Even if we don't get hurt ourselves, it may still deeply affect us, and hurt us emotionally, if a family member, an acquaintance, or a fellow employee is injured or killed. It even stirs up feelings of sympathy in us when we hear about a serious electrical injury or fatality that happened to a stranger. For these reasons, we should work safely ourselves, look out for the safety of those around us, and help educate as many people as we can about electrical safety.

7.4.2 Business considerations

The second reason for practicing electrical safety is the fact that safety makes good business sense. The "bottom line" of most businesses is to keep costs down and earn profits. If serious electrical injury, death, or property damage occurs at a particular company facility as a result of unsafe acts or unsafe conditions, it costs that company large amounts of money and aggravation in both tangible and intangible ways. Consider the possible costs of the following items:

a) Time lost by the injured employee;
b) Medical bills;
c) Time lost by supervisors and managers trying to find out what really happened;
d) Time lost and aggravation caused by answering many questions from all directions;
e) Time lost by supervisors and managers writing reports;
f) Time lost by other employees sympathizing and discussing the accident;
g) Lost production quantity or quality;
h) Loss of test and research data if unexpected electrical outages result;
i) Possible increase in insurance rates;
j) Possibility of lawsuits by the injured, or the family, claiming negligence;
k) Possible intervention and fines by a law enforcing agency such as OSHA;
l) Payment for any necessary outside consultants;
m) Time lost escorting investigators;
n) Discomfort of explaining "why" to the family of the victim;
o) Lowered morale of fellow employees if employer is thought to be negligent;
p) Replacement or repair costs of damaged equipment;
q) General disruption of the normal operating routine;
r) Mistrust by customers of a company with a poor safety record.

Some of the above costs are difficult to put a value on, but they are real. The cost of one serious incident could be more than the cost of establishing and maintaining a good electrical safety program for several years. In the long run, the safe way is the least expensive way to run a business.

7.4.3 Regulatory requirements and legal considerations

The third and most compelling reason for practicing electrical safety is the fact that laws exist that mandate certain requirements for electrical safety and fire protection. Breaking these laws could mean fines for companies and possibly even imprisonment for management personnel.

In the U.S., there are two main documents relating to electrical safety for use by governing agencies: the Occupational Safety and Health Administration (OSHA) Standards (see 7.4.3.1) and the National Electric Code® (NEC®) (NFPA 70-1996) (see 7.4.3.2).

There are also other recognized standards and guides that contain electrical safety information that might be referenced in a legal case.

7.4.3.1 OSHA regulations

The U.S. Department of Labor has written laws under Title 29 of the Code of Federal Regulations (29 CFR) that establish requirements for electrical installations and electrical safe practices in that country. These are commonly called the OSHA Regulations.[2] Part number 1910 covers the requirements for general industry, and Part number 1926 covers the requirements for the construction industry.

NOTE—Throughout the rest of this guide, OSHA regulation numbers will be referred to by abbreviation. For example, definitions for the electrical subpart are contained in 29 CFR 1910.399.

The subparts of Part 1910 that are of particular interest relative to electrical safety are

— Subpart S—Electrical, all sections;
— Subpart R—Special Industries, Section 1910.269—Electric Power Generation, Transmission, and Distribution;
— Subpart I—Personal Protective Equipment, Section 1910.137—Electrical Protective Devices;
— Subpart J—General Environmental Controls, Section 1910.147—Hazardous Energy Control (Lockout/Tagout).

The Subparts of Part 1926 that are of particular interest relative to electrical safety are

— Subpart K—Electrical, all sections;
— Subpart V—Power Transmission and Distribution, all sections.

[2]OSHA regulations are published in the *Federal Register.* The *Federal Register* is available from the Superintendent of Documents, U.S. Government Printing Office, Washington, DC 20402 (telephone 202-783-3238) on a subscription or individual copy basis.

There are many other subparts in both 29 CFR 1910 and 29 CFR 1926 that refer to topics that are not directly electrical safety in nature, but which are applicable to electrical work. It would be wise to review the OSHA Table of Contents of these two parts.

Although it is not an easy task to train everyone, all employees are required to abide by these laws, and any electrical safety program in the U.S. shall be based on these laws. Other countries of the world may have their own laws on which their electrical safety programs should be based. In any case, before beginning electrical work anywhere, make sure the applicable laws regarding electrical safety are known and followed.

7.4.3.2 The National Electrical Code®

NFPA 70-1996, commonly called the NEC, is well known as an electrical fire protection and safety document related to the installation of "premises" wiring. As defined in the NEC, premises wiring is "that interior and exterior wiring, including power, lighting, control, and signal circuit wiring together with all of their associated hardware, fittings, and wiring devices, both permanently and temporarily installed, that extends from the service point of utility conductors or source of a separately derived system to the outlet(s)." In itself, the NEC is a standard of advisory information offered for use in law and for regulatory purposes. It can be, and is, adopted and made mandatory by many governing agencies in the U.S. and other countries, thereby giving it the force of law in those jurisdictions. It should also be remembered, however, that the NEC is a minimal standard; therefore, its requirements sometimes have to be exceeded to meet functional necessities, good engineering judgment, and enhanced safety practices. The NEC will be referenced many times in this guide.

7.4.3.3 Other standards

Another electrical safety standard published by the NFPA is NFPA 70E-1995. When the OSHA electrical standards were developed in 1969 and 1970, they were based on the NEC. As OSHA focused more on all aspects of electrical safety, the need was created for a consensus document to assist them in preparing electrical safety requirements. The first edition of this document was published in 1979 and was most recently updated in 1995. Although this document does not yet have the same extensive recognition as the NEC, it does provide the latest thinking on the subject of electrical safety, particularly in the area of safe practices. Many parts of the current OSHA regulations 29 CFR 1910, Subpart S, were derived from NFPA 70E-1995.

Another good document that contains electrical safety information is the National Electrical Safety Code® (NESC®) (Accredited Standards Committee C2-1997). This standard applies primarily to outdoor electrical transmission, distribution, and communication systems, equipment, and associated work practices; as opposed to the NEC which concerns itself primarily with premises wiring. The 1997 NESC contains five parts:

a) The first part has no number but contains Sections 1, 2, 3, and 9, and covers the Introduction, Definitions of Special Terms, References, and Grounding Methods for Electric Supply and Communication Facilities.

b) Part 1 includes Sections 10–19 and covers Rules for Installation and Maintenance of Electric supply Stations and Equipment.
c) Part 2 includes Sections 20–28 and covers Safety Rules for Installation and Maintenance of Overhead Electric Supply and Communication Lines.
d) Part 3 includes Sections 30–39 and covers Safety Rules for Installation and Maintenance of Underground Electric Supply and Communication Lines.
e) Part 4 includes Sections 40–44 and covers Rules for the Operation of Electric Supply and Communication Lines and Equipment.

NFPA 70B-1998 is a document whose purpose is to reduce the hazards to life and property that can result from the failure or malfunction of industrial-type electrical systems and equipment. Along with its maintenance guidance, it also addresses electrical safety.

The National Electrical Manufacturers Association (NEMA) has many standards on electrical products and systems. Their product standards have often served as a basis for Underwriter Laboratories' (UL) safety standards. Both NEMA and UL standards are formulated on a consensus basis and should be considered as minimal requirements.

The Color Book Series by the Institute of Electrical and Electronics Engineers (IEEE) provides recommended practices that go beyond the minimal requirements of the NFPA, NEMA, and UL standards. When designing electrical power systems for industrial and commercial facilities, consideration should be given to the design and safety requirements of the following IEEE color books:

— IEEE Std 141-1993, IEEE Recommended Practice for Electric Power Distribution for Industrial Plants (*IEEE Red Book*) [B9];
— IEEE Std 241-1990, IEEE Recommended Practice for Electric Power Systems in Commercial Buildings (*IEEE Gray Book*) [B10];
— IEEE Std 242-1986, IEEE Recommended Practice for Protection and Coordination of Industrial and Commercial Power Systems (*IEEE Buff Book*) [B11];
— IEEE Std 399-1997, IEEE Recommended Practice for Power Systems Analysis (*IEEE Brown Book*) [B12].

It should be noted that, included under the first principles in the IEEE Buff Book, are the statements, "safety has priority over service continuity, equipment damage, or economics" and "engineers engaged in the design and operation of electrical system protection should familiarize themselves with the most recent OSHA regulations and all other applicable regulations relating to human safety." This often means going beyond minimum code requirements to assure adequate safety.

7.5 Summary

In summary, electrical hazards exist, and will continue to exist, in our world of ever-increasing technology. In order to prevent injury and loss of life, everyone should learn to recognize the electrical hazards associated with their respective workplaces and job assignments, and know how to take appropriate precautions to avoid injury. This chapter has covered the kinds of injuries that can occur and some true examples of incidents. Remembering these incidents, and telling friends and associates about them, may instill in someone the awareness and discipline that could someday save a life.

There are a lot of reasons why everyone should receive some level of electrical safety training. Familiarity with the OSHA regulations, the NFPA electrical codes and standards, and the IEEE standards can increase safety awareness for all electrical employees, and perhaps may save an employee or friend from injury or death.

7.6 References

This chapter shall be used in conjunction with the following publications. If the following publications are superseded by an approved revision, the revision shall apply.

Accredited Standards Committee C2-1997, National Electrical Safety Code® (NESC®).[3]

CFR 29 Part 1910, Occupational Safety and Health Standards.[4]

CFR 29 Part 1926, Safety and Health Regulations for Construction.

NFPA 70-1996, National Electrical Code® (NEC®).[5]

NFPA 70B-1998, Recommended Practice for Electrical Equipment Maintenance.[6]

NFPA 70E-1995, Standard for Electrical Safety Requirements for Employee Workplaces.

7.7 Bibliography

Additional information may be found in the following sources.

[B1] Andrews, J. J., McClung, L. B., and White, T. J., "Ensuring that electrical equipment is safe for its intended use," *IEEE Industrial and Commercial Power Systems Technical Conference,* May 1998.[7]

[3]The NESC is available from the Institute of Electrical and Electronics Engineers, 445 Hoes Lane, P.O. Box 1331, Piscataway, NJ 08855-1331, USA (http://www.standards.ieee.org/).

[4]CFR publications are available from the Superintendent of Documents, U.S. Government Printing Office, P.O. Box 37082, Washington, DC 20013-7082, USA.

[5]The NEC is available from Publications Sales, National Fire Protection Association, 1 Batterymarch Park, P.O. Box 9101, Quincy, MA 02269-9101, USA (http://www.nfpa.org/). Copies are also available from the Institute of Electrical and Electronics Engineers, 445 Hoes Lane, P.O. Box 1331, Piscataway, NJ 08855-1331, USA (http://www.standards.ieee.org/).

[6]NFPA publications are available from Publications Sales, National Fire Protection Association, 1 Batterymarch Park, P.O. Box 9101, Quincy, MA 02269-9101, USA (http://www.nfpa.org/).

[7]IEEE publications are available from the Institute of Electrical and Electronics Engineers, 445 Hoes Lane, P.O. Box 1331, Piscataway, NJ 08855-1331, USA (http://www.standards.ieee.org/).

[B2] Castenschiold, R., and Love, D. J., "Electro-forensic engineering—an emerging profession," *IEEE Industry Applications Magazine,* Mar./Apr. 1997.

[B3] Dalziel, C. F., "Electrical shock hazard," *IEEE Spectrum,* pp. 41–50, Feb. 1972.

[B4] Doughty, R. L., Epperly, R. A., and Jones, R. A., "Maintaining safe electrical work practices in a competitive environment," *IEEE Transactions on Industry Applications,* vol. 28, no. 1, Jan./Feb. 1992.

[B5] Drouet, M. G., and Nadeau, F., "Pressure waves due to arcing faults in a substation," *IEEE Transactions on Power Apparatus and Systems,* vol. PAS-98, pp. 1632–1635, 1979.

[B6] Gallagher, J. M., and McClung, L. B., "Electrical hot work safety program," *IEEE Transactions on Industry Applications,* vol. 24, no. 1, Jan./Feb. 1988.

[B7] Goldberg, D. L., and Castenschiold, R., "The designer of industrial and commercial power systems and code interpretations," *IEEE Transactions on Industry Applications,* vol. 29, no. 5, Sept./Oct. 1993.

[B8] Greenwald, E. K., ed., *Electrical hazards and accidents.* Van Nostrand Reinhold, 1991

[B9] IEEE Std 141-1993, IEEE Recommended Practice for Electric Power Distribution for Industrial Plants (*IEEE Red Book*).

[B10] IEEE Std 241-1990 (Reaff 1997), IEEE Recommended Practice for Electrical Power Systems in Commercial Buildings (*IEEE Gray Book*).

[B11] IEEE Std 242-1986 (Reaff 1991), IEEE Recommended Practice for Protection and Coordination of Industrial and Commercial Power Systems (*IEEE Buff Book*).

[B12] IEEE Std 399-1997, IEEE Recommended Practice for Power Systems Analysis (*IEEE Brown Book*).

[B13] Kouwenhoven, W. B., and Milnor, W. R., "Field treatment of electric shock cases," *AIEE Transactions on Power Apparatus and Systems,* vol. 76, pp. 82–84, Apr. 1957; [discussion of: pp. 84–87].

[B14] Lee, R. H., "Electrical safety in industrial plants," *IEEE Spectrum,* June 1971.

[B15] Lee, R. H., "Pressures developed by arcs," *IEEE Transactions on Industry Applications,* vol. IA-23, no. 4, p. 760, July/Aug. 1987.

[B16] Lee, R. H., "The other electrical hazard: Electrical arc blast burns," *IEEE Transactions on Industrial Applications,* vol. 1A-18, no. 3, p. 246, May/June 1982.

Chapter 8
Establishing an electrical safety program

8.1 General discussion

Before a sports team goes into any game, the coaches prepare a game plan designed to beat the opponent. If all members of the team perform according to the plan, the team has a much better chance of winning. An electrical safety program is much like that game plan. It is a plan designed so that neither workplace conditions, nor the actions of people, expose personnel unnecessarily to electrical hazards. Establishing an electrical safety program, and making sure that employees follow it, can mean "winning the game" against accidental injury or death due to electrical incidents.

For all of the reasons mentioned in 7.4, an employer should develop and implement some form of an electrical safety program to give overall safety directions for facility activities related to electrical work.

8.1.1 Acronyms and abbreviations

AHJ	authority having jurisdiction
CPR	cardiopulmonary resuscitation
NEC	National Electrical Code®
NFPA	National Fire Protection Association
OSHA	Occupational Safety and Health Administration

8.2 Purpose

An electrical safety program has the following five objectives:

a) To make personnel aware that there are rules, responsibilities, and procedures for working safely in an electrical environment;

b) To demonstrate the employer's intention to fully comply with federal law;

c) To document general requirements and guidelines for providing workplace facilities that are free from unauthorized exposure to electrical hazards;

d) To document general requirements and guidelines that direct the activities of personnel who could be deliberately, or accidentally, exposed to electrical hazards;

e) To encourage, and make it easier for, each employee to be responsible for his or her own electrical safety self-discipline.

If the safety program is formal, and documented in writing, an employer can comply effectively with the regulatory requirements mentioned in Chapter 7 and prove such compliance when requested to do so.

8.3 Scope

The scope of the electrical safety program should address the needs of all employees, as well as contractors and visitors, within a company or at a facility. Everyone should be aware that the program exists, and should be very familiar with the parts that pertain to his or her own particular job assignments. The program should address policy, requirements, responsibilities, and guidance in general terms. Specifics and details could be placed in other subdocuments, such as procedures, to which the program can simply refer.

The size of the program depends upon the size of the company, both in the number and complexity of facilities, and the number of personnel involved with electrical work. The guidance in this chapter gives the larger picture so that companies can give thorough consideration to their own specific needs. The program should be kept as simple and as easy to understand as possible. At the same time, however, it should cover all the needs of the company.

8.4 Content of program

As mentioned in Chapter 7, any electrical safety program shall be established with consideration of its legal implications. In the U.S., the primary legal concern is the Occupational Safety and Health Administration (OSHA) regulations. Other countries also have laws on which programs are based. It is important that the laws of a country are known and understood before the electrical safety program is defined and implemented.

A complete electrical safety program should contain directives on the following subjects:

a) Management commitment (8.4.1);
b) Organizational support (8.4.2);
c) Electrical safety policy (8.4.3)
    1) Electrically safe facilities;
    2) Documented safe electrical work practices;
d) Training and qualification of all personnel (8.4.4);
e) Use of protective equipment, tools, and protective methods (8.4.5);
f) Use of electrical equipment (8.4.6);
g) Documentation (8.4.7);
h) Oversight and auditing (8.4.8);
i) Technical support (8.4.9);
j) Emergency preparedness (8.4.10).

These items are discussed in more detail in 8.4.1 through 8.4.10.

8.4.1 Management commitment

An electrical safety program may be totally ineffective if it is not strongly supported at the highest management levels. Not only should management support the program, but management should also ensure that it is truly being implemented in the workplace. Otherwise, the program may become another pile of paper in a manual somewhere. Management should not just delegate electrical safety responsibility, but should show genuine interest at all management levels. Management should believe that there is a real value in an electrical safety program, both from humanitarian and financial standpoints. (Refer to 7.4.) Just as a coach will state his game plan, enforce it, then reinforce it, so too should the electrical safety program be stated, enforced, and reinforced by management.

As part of the program, management should first establish an electrical safety policy and identify the line organization(s) to implement that policy. Management should also direct that the program be audited and improved periodically. Management should demonstrate by its commitment that safety is truly a top priority in business. Safety should be given at least equal, and maybe even greater, importance than production, cost, quality, and morale. In demonstrating this importance, management should be consistent when it comes to applying this equality by not lowering safety standards when confronted with other business pressures.

8.4.2 Organizational support

In order to accomplish a company's business objective, management structures an organization. An electrical organization and a safety organization are usually included in the business objective. One or both of these organizations should be designated to accomplish the electrical safety objective. To accomplish the electrical safety objective, there need to be individuals or groups to perform the following functions:

— Management;
— Design;
— Installation;
— Facility operations;
— Maintenance;
— Training;
— Purchasing;
— Visitor and contractor liaison;
— General industrial safety;
— Electrical safety authority.

Every organization has a hierarchy of authority. The National Fire Protection Association (NFPA) defines the authority having jurisdiction (AHJ) as the organization, office, or individual responsible for "approving" equipment, an installation, or a procedure. The definition is followed by a note that gives examples of who this authority might be. In some jurisdictions, this authority is easily identified. In many large organizations, however, the managers who have ultimate authority usually do not have the expertise themselves to make decisions in specific technical areas. They usually delegate that authority or, at least, get advice from specialists before making final decisions. The function of the AHJ, therefore, usually rests at a lower level in the organization.

Electrical safety processes are the same as other safety processes, with only the energy source being different. Consequently, the electrical safety program can blend easily with any existing overall safety program and organization. Company management should designate some part of the organization to be the focal point, or the authority, for electrical safety. Depending on the size of the company, this focal point might be a department, a committee, or an individual or two. This designee should have the responsibility of knowing the electrical safety regulations and standards, as well as taking care of the following functions:

- Taking ownership of the electrical safety program;
- Developing and revising company electrical standards;
- Providing interpretations of nationally recognized codes and standards;
- Providing guidance for facility configuration management;
- Resolving National Electrical Code® (NEC®) (NFPA 70-1996) [B2][1] inspection questions;
- Establishing and documenting good safe work practices;
- Providing technical input for OSHA interpretations;
- Providing guidance for electrical training programs;
- Providing guidance for procedure preparation;
- Providing consultation services to management;
- Reviewing electrical safety incidents and participating in investigations;
- Issuing summaries and lessons-learned about electrical safety incidents;
- Evaluating nonlisted electrical equipment or knowing how to get an evaluation done.

These tasks can be divided into two broad categories for those cases in which the responsibilities would be too much for one group. The first is design and installation concerns, and the second is operational and maintenance concerns.

8.4.3 Electrical safety policy

Management's electrical safety policy may be stated in a policy manual or it could be stated in the electrical safety program. It might simply be stated in the following manner: "Electrically safe facilities shall be established and maintained. All work involving electrical energy shall be performed in a safe manner." Unfortunately, these words are sometimes the extent of management's direction. If management stops here without providing more thought and resources to carry the policy through to implementation, the resulting electrical safety performance in the field will be poor.

[1]The numbers in brackets correspond to those of the bibliography in 8.6.

An important basic rule that should be derived from the policy statement is that work on or near any exposed energized electrical conductors or circuit parts should be prohibited, except under justified, controlled, and approved circumstances. This is the safest policy. Of course, everyone in the electrical business knows that there are a lot of reasons why someone would like to take exception to that policy. For example, taking a voltage measurement is a very common and necessary task in the electrical business that involves exposure to energized conductors. As indicated by some of the cases in Chapter 7, however, things can go wrong even during such a common task. Therefore, taking exceptions to the basic rule should be strictly controlled. Guidelines for what constitutes justification are given in NFPA 70E-1995,[2] Part II, Section 2-3, where it states, "Exposed energized electrical conductors or circuit parts to which an employee might be exposed shall be put into an electrically safe work condition before an employee works on or near them, unless the employer can demonstrate that de-energizing introduces additional or increased hazards or is infeasible due to equipment design or operational limitations." This statement is further explained in the notes of that section.

8.4.3.1 Providing and maintaining electrically safe facilities

The first statement of the electrical safety policy states that "electrically safe facilities shall be established and maintained." That statement covers the following concerns, which should be detailed in various company documents:

a) Design;
b) Installation;
c) Inspection;
d) Maintenance (breakdown and preventive);
e) Standards;
f) Safety audits of workplace conditions;
g) Organizational structures for tasks a) through f);
h) Technical training and qualifications of personnel;
i) A technical authority to respond to questions or concerns about design and installation.

It is important that facilities and systems be initially designed and installed in such a manner as to provide for all personnel a safe workplace, free from exposure to electrical hazards. Complying with established federal, state, and local laws and codes helps to ensure safe installations. After facility start-up, this safe environment shall be maintained as close as possible to the initial condition. Following the guidance of nationally recognized standards can help to ensure safe facilities. As mentioned in Chapter 7, some of the better known laws, codes, standards, and guides include those of OSHA, the NFPA, and the IEEE.

Electrical equipment integrity is a fundamental part of an electrical safety program. Particular emphasis should be placed on the integrity of enclosures, insulation, grounding, and circuit protective devices. The main point to be understood is that exposure to hazards can be minimized by proper design, installation, and maintenance of equipment.

---

[2]Information on references can be found in 8.5.

Good facility standards and drawings are also important elements in establishing and maintaining a safe facility. Chapter 9 delves into the subject of electrically safe facilities in greater detail.

8.4.3.2 Implementing safe electrical work practices

The second part of the electrical safety policy states that "all work involving electrical energy shall be performed in a safe manner." This statement would include such things as

a) General policy about avoiding work on energized electrical equipment;
b) Work authorization;
c) Standards;
d) Use of safe practices and procedures;
e) Oversight groups;
f) Safety training and qualifications of personnel;
g) Safety audits and self-assessments of personnel activities;
h) A technical authority to respond to questions or concerns regarding safe work practices in operations and maintenance.

Again, these concerns should be written into official company documentation.

Safe electrical work practices are perhaps the most important part of the whole electrical safety program. It is a fact that most injuries and fatalities are a result of the actions of people as opposed to workplace conditions. Electrical safety principles (see Whelan and Floyd [B3]) should be identified and taught to employees. These principles are

— Plan every job.
— Anticipate unexpected events.
— Use the right tool for the job.
— Use procedures as tools.
— Isolate the equipment.
— Identify the hazard.
— Minimize the hazard.
— Protect the person.
— Assess people's abilities.
— Audit these principles.

Establish administrative controls in the form of policy, required procedures, and guidelines to direct the activities of personnel whose assignments are such that they could very easily become exposed to electrical hazards. Before any work is performed, there should be a hazard/risk analysis, work authorization, and a job-briefing that focuses heavily on safety. See Chapter 10 for details on this important subject.

8.4.4 Training and qualification of all personnel

The electrical safety program should direct that all personnel have electrical safety training appropriate to their assigned jobs. This requirement then would include some level of training for every employee. It also includes ensuring that any contractors or visitors who might become exposed to electrical hazards have been trained and are qualified in electrical safety by their own employers. If they have not been trained, they should not be exposed to such hazards at the facility until the training is completed.

The best way to help people avoid hurting themselves is to train them. In this day and age, because they are surrounded by electrical equipment, almost every person, including the so-called unqualified person, utilizes electrical energy in one form or another. It is, of course, impractical and unnecessary to train all employees to be electrical experts. It is important, however, to train them to know how to recognize where electrical hazards might be located both in the workplace and at home. At the very least, they should be trained on, and qualified for, the equipment that they use, and should know how to recognize when something is going wrong with any electrical components in their vicinity. They should also be trained to understand what they are not allowed to do.

A document should be created as part of a company's employee safety manual that covers the basic electrical safety awareness that is applicable to all employees. The document should be used as a tool to help understand hazards and how to deal with them. This document should contain, in a general sense, the following:

a) Definitions of electrical safety-related terms;
b) Electrical safety rules that are applicable to all personnel;
c) Guidance to the location of more detailed rules for electrical personnel;
d) Discussion on the use of electrical equipment;
e) Overview of personal protection requirements near electrical hazards;
f) Training requirements to become more electrically qualified.

Of course, those persons whose jobs are more directly related to the electrical business need a much deeper and more specific level of training than just the fundamental knowledge of electricity. They should have electrical safety training that covers all of the equipment on which they might work, safe practices and methods, and the use and care of personal protective equipment, tools, and test equipment. Training in some emergency response procedures is also important (see 8.4.10).

8.4.5 Use of protective equipment, tools, and protective methods

No matter how good the workplace conditions are, and no matter how well trained personnel are, somehow or other, things occasionally go wrong when least expected. So, in addition to a good workplace environment and safe practice controls, it is wise to use other protective measures "just in case." Every job should be evaluated to determine what protective equipment and tools would lessen the risk of injury. Protective methods are the little extra safety precautions that can be taken to reduce personnel exposure to hazards. Chapter 11 gives details about some of the additional ways employees can be protected against the unexpected.

8.4.6 Use of electrical equipment

People are not only surrounded by electrically powered equipment, but they also operate and use the equipment. There is fixed equipment such as switchgear, control panels, and wall switches; and there is portable equipment such as power tools, extension cords, and test equipment. It is also important to be aware of the environment in which the equipment is being used. Direction should be provided for the proper use of such equipment. Chapter 12 provides detailed information about the use of electrical equipment.

8.4.7 Documentation

There is a fair amount of documentation that is required in order to really have an effective electrical safety program. As is mentioned throughout this book, these documents include work authorization, standards, procedures, guidelines, drawings, and equipment records. Instead of discussing the various documents in detail in this chapter, the documents are discussed in the appropriate chapter, along with the subject matter to which the document applies.

A good document management system is very important to the safe operation and maintenance of a facility. Outdated or erroneous documents not only cause confusion and delays, but they also can cause electrical-related safety incidents. The distribution of documents should be controlled so that only the currently applicable version is readily accessible. Documents should also be located in the places in which they are needed. Otherwise, users may bypass them or delays may be caused while users try to find them. More information on the subject of procedures and documentation can be found in Chapter 10.

8.4.8 Oversight and auditing

Periodically, a company should perform a self-assessment to determine how well their written electrical safety program is actually being implemented. To be of value, the assessment should be very objective, without trying to blame people. The goal is to improve the safety performance, not to punish employees. Besides, if the program is not being carried out effectively, a significant part of the blame belongs to management. Occasionally, it is also prudent to have the company's safety organization, or even an outside-contracted safety auditor, perform an electrical safety audit. This way, a set of eyes that is not so familiar with the facility, and does not have a fear of retaliation from local management, can discover things that self-assessors might overlook.

The electrical safety program should contain a requirement that the written program itself be occasionally audited. The auditing should be designed to identify new or revised requirements, as well as weaknesses in the system. Attention should be payed not only to the written work practices and procedural requirements, but also to how well personnel appear to understand and implement these requirements.

Whenever electrical safety-related incidents or accidents occur, there should be a thorough investigation to determine the root cause and contributing factors. As a result of the investigation, actions should be taken to prevent a recurrence of the incident. Lessons-learned information should be distributed to all personnel who could get involved in, or influence decisions about, similar future situations.

8.4.9 Technical support

A company should have easy access to a qualified engineering organization and/or qualified consultants. These persons could be either in-house or contracted employees. This type of personnel not only can provide designs for new facilities and equipment, but also can provide guidance for many of the daily or recurring problems that interfere with a safe, smooth-running operation. Quite a few of these engineers and consultants stay abreast of the latest developments in their fields of expertise by belonging to nationally recognized professional organizations, such as the IEEE and the NFPA. These engineers can provide guidance on all aspects of operations, maintenance, and safety because, if they don't already know the answers, they have a network to help them find the answers. Companies should encourage, and strongly support, their in-house engineers to join nationally recognized professional organizations related to their fields of endeavor.

8.4.10 Emergency preparedness

Despite all of our efforts to have an injury-free workplace, the unexpected occasionally happens. One of the items that is often forgotten when preparing an electrical safety program is the need for emergency preparedness. Being able to respond quickly to an electrical shock or burn injury could be the difference between the life and death of the victim. All electrical workers should be taught first aid and cardiopulmonary resuscitation (CPR). There should also be periodic refresher training on these subjects. If it is necessary to work on or near energized conductors, it is recommended that an electrical-safety qualified attendant be present to provide assistance if needed. Rescue equipment should be available in case a worker gets "hung-up" on an energized conductor.

The location and phone number(s) of the nearest qualified medical assistance should be known. A phone or some other communication method should be available at every job location that involves work on energized conductors, or has other elements of risk. Emergency responders should be taught basic electrical safety so that they don't get themselves hurt while trying to help a person who is injured in an electrical incident.

Try to anticipate the unexpected injury possibilities of a job and have an emergency plan ready.

8.5 References

This chapter shall be used in conjunction with the following publications. If the following publications are superseded by an approved revision, the revision shall apply.

NFPA Regulations Governing Committee Projects, 1997 NFPA Directory, Section 2-3.6.1.[3]

NFPA 70E-1995, Standard for Electrical Safety Requirements for Employee Workplaces.

8.6 Bibliography

Additional information may be found in the following sources:

[B1] *Department of Energy Model Electrical Safety Program*, United States Department of Energy.[4]

[B2] NFPA 70-1996, National Electrical Code® (NEC®).[5]

[B3] Whelan, C. D., and Floyd, H. L., "Basics for electricians," *Plant Services Magazine*, Dec. 1992.

[3]NFPA publications are available from Publications Sales, National Fire Protection Association, 1 Batterymarch Park, P.O. Box 9101, Quincy, MA 02269-9101, USA (http://www.nfpa.org/).

[4]Department of Energy publications are available from the Superintendent of Documents, U.S. Government Printing Office, P.O. Box 37082, Washington, DC 20013-7082, USA.

[5]The NEC is available from Publications Sales, National Fire Protection Association, 1 Batterymarch Park, P.O. Box 9101, Quincy, MA 02269-9101, USA (http://www.nfpa.org/). Copies are also available from the Institute of Electrical and Electronic Engineers, 445 Hoes Lane, P.O. Box 1331, Piscataway, NJ 08855-1331, USA (http://www.standards.ieee.org/).

Chapter 9
Providing and maintaining electrically safe facilities

9.1 General discussion

As was mentioned in Chapter 8, there are five objectives of the electrical safety program. This chapter discusses the objective of providing workplace facilities that are safe from exposure to electrical hazards.

When electrical installations are properly designed, installed, and maintained, workplace conditions should be such that they will not be the cause of an electrical shock, burn, or blast. Well-designed enclosures and proper clearances protect employees under normal and most abnormal operating conditions. Proper maintenance of electrical equipment can reestablish the initial safe condition as the equipment ages. It should always be remembered, however, that when it is necessary for personnel to enter electrical equipment enclosures, many of the protective barriers that the design and initial installation have provided may have to be removed. In these cases, electrical safe practices that are appropriate to the task should be used. (See Chapter 10.)

The two major documents that carry the force of law relative to electrical design, installation, and maintenance requirements in the U.S. are extremely important toward establishing an electrically safe workplace. As was mentioned in Chapter 7, the first is the Occupational Safety and Health Administration (OSHA) regulations; in particular, 29 CFR 1910,[1] Subpart S, pertaining to general industry; 29 CFR 1910, Subpart R, pertaining to special industries; and 29 CFR 1926, Subpart K, pertaining to the construction industry. There are also some other subparts that contain requirements related to electrical work. The second major document is the National Electrical Code® (NEC®) (NFPA 70-1996), which carries the force of law only when it is adopted by a governing entity. Practically all governing bodies in the U.S., and even some other countries, have adopted the NEC.

There are many other good references that cover safe design and installation of electrical equipment and systems. It would be difficult to mention them all. Included among them, however, are the *IEEE Color Book* series (of which this book is one), NFPA 70E-1995, and the National Electrical Safety Code® (NESC®) (Accredited Standards Committee C2-1997).

NFPA 70E-1995 was created to serve the needs of OSHA. Part I of NFPA 70E-1995 has specifically identified those parts of the NEC that are directly associated with employee safety in the workplace. It duplicates those parts of the NEC that can be safely inspected while electrical equipment is energized. Although the information in NFPA 70E-1995 cannot be used alone in lieu of other requirements of the NEC, NFPA 70E-1995 does highlight design and installation considerations that are safety-related. In addition, NFPA 70E-1995 contains Part II, Safety-Related Work Practices; Part III, Safety-Related Maintenance Requirements; and Part IV, Special Requirements.

[1]Information on references can be found in 9.7.

It is not the intention of this chapter to repeat, in detail, the safety-related design, installation, and maintenance information that is already found in other recognized regulations, codes, and standards. This chapter will, however, give an overview of how to find and use the helpful information found in those documents.

9.1.1 Acronyms and abbreviations

NEC National Electrical Code®
NESC National Electrical Safety Code®
NRTL nationally recognized testing laboratory
OSHA Occupational Safety and Health Administration

9.2 Design considerations

Providing safe workplace facilities begins with the initial design of a facility, process, or piece of equipment. The design process begins with contacting the customer and thoroughly understanding the customer's objectives for his facility, including, in particular, the safety concerns. The documents mentioned in Chapter 8 and in 9.1 contain requirements for design. These documents, however, are not design specifications, and should be used with care only by qualified personnel.

The NEC is not a design manual for untrained persons. The requirements it contains are often the minimum that are essential to achieve fire protection and safety for the facilities. Many times, however, design specifications need to be even more strict in order to achieve a cost-effective, well-functioning installation. That is why the NEC is intended to be used only by the following kinds of qualified people:

a) Qualified engineers and designers in the design of electrical equipment and facilities;
b) Electrical inspectors and others who exercise legal jurisdiction;
c) Insurance inspectors to ensure a safe, hazard-free installation;
d) Electrical contractors and electricians to meet installation requirements;
e) Instructors to educate and refresh electrical personnel.

The NEC contains many requirements to help a design organization develop a safe facility from an electrical standpoint. These requirements cover the following topics:

— Wiring and protection;
— Methods and materials;
— Equipment for general use;
— Special occupancies;
— Special equipment;
— Special conditions;
— Communication systems.

For safety purposes, design should take into account the following items:

- Protection against shock, burn, and blast;
- Fire protection;
- Illumination;
- Working space;
- Drawings;
- Equipment identification;
- Grounding and bonding.

9.2.1 Design for shock, burn, and blast protection

Meeting NEC requirements for electrical equipment installations is only one part of a safe design. The design of electrical equipment should also include extra features that can reduce the possibility of equipment damage and personnel injury. For example, arcing faults can be very dangerous and destructive. Considerations to reduce the consequences of such faults should be addressed at the design stage.

Power distribution equipment should be designed so that it is easily operated and maintained. It should also be lockable.

Short-circuit analysis or fault-current analysis is very important toward identifying the fault current available at any given point in a system. This is important in the hazard/risk analysis (see Chapter 10) to determine the possible consequences of employee exposure to an electrical hazard.

The proper grounding of equipment that utilizes electrical energy is essential to protection from electrical shock. The subject of grounding is covered in Article 250 of the NEC.

In an industrial environment, electrical equipment should be heavy duty in order to endure the more frequent usage and higher available fault energy. In the U.S., the electrical equipment that is selected should also have been tested and listed by a nationally recognized testing laboratory (NRTL). If listed equipment is not available to perform a particular function, some jurisdictions would require an NRTL field evaluation of the nonlisted equipment. If not otherwise required, the facility should perform its own safety evaluation.

The environment in which electrical equipment is installed should also be taken into account during design. The deteriorating effects of weather and chemicals can eventually compromise equipment safety features.

9.2.2 Design for fire protection

In addition to providing designed safety against shock, burn, and blast, the workplace should also be designed for fire protection. Fires occasionally develop in electrical equipment due to loose connections or overload conditions that may not be cleared properly by protective devices. Protection against fire is a good reason to ensure proper design, installation, and maintenance. Selecting equipment that is tested and listed by an NRTL, as mentioned in the previous paragraph, helps to reduce the risk of fire in the equipment.

In addition to the design of the electrical system and equipment themselves, there should be fire detection and suppression equipment permanently installed or readily accessible in the facility around or near the electrical equipment. Such equipment could possibly include smoke detectors, sprinkler systems, and portable fire extinguishers. Also, the workplace should be designed so that escape routes are sufficiently wide, clear of obstructions, and well marked. Proper illumination of the paths of egress by normal lighting, emergency lighting, and exit signs is also very important. For more about fire safety, refer to other NFPA standards.

In order to avoid fire and explosions, pay attention to electrical installations in classified hazardous areas in accordance with Chapter 5 of the NEC.

The proper sizing of conductors is important in order to protect against the overheating of the conductors and their surroundings. Article 310 of the NEC covers this subject.

Overcurrent protection is important to avoid equipment burndowns when overloads or faults occur. Short-circuit analysis is a major part of establishing adequate fault protection. Overcurrent protection is addressed in Article 240 of the NEC.

Motors and generators need to be protected against overheating so that they do not cause a fire. Article 430 of the NEC covers motors. Article 445 of the NEC covers generators.

9.2.3 Illumination

Lighting systems in the area around electrical equipment should be designed to provide adequate illumination of the vertical surfaces of the electrical equipment. Proper illumination is not only important for performing normal tasks, but it is also very important from a safety standpoint when installations are being checked-out at start-up, as well as during maintenance and troubleshooting after turnover to operations. Poor illumination might be the cause of an electrician putting a tool in the wrong place and creating a flash. In a high-energy circuit, this mistake could be fatal. If permanently installed lighting is felt to be inadequate, no work should be attempted in any area that might contain a hazard. Some temporary lighting equipment should be obtained and used at the work location. Perhaps a suggestion should be made to improve the lighting in that area. Emergency lighting should also be installed to provide necessary illumination in order to avoid safety problems during a sudden power outage.

9.2.4 Work spaces and working clearances

Electrical equipment should be designed with adequate working spaces both within the equipment and around it. Installations in cramped spaces or areas where accessibility is difficult should be avoided. Working on or near exposed energized electrical parts in cramped spaces is especially dangerous because reflex reactions, from bumping into objects, could cause a person to involuntarily contact the energized parts and receive a shock or instigate a flash. Dimensions for access to, and clearances around, electrical equipment are given in Article 110 of the NEC.

Allow room for future expansion and rearrangement in the initial designs. Although it is usually not possible to predict the future, more often than not, someone will want growth or improvement. Allowing extra space initially helps to avoid the temptation to squeeze things in and possibly create a working-space safety problem in the future.

9.2.5 Drawings

Drawings should be created initially not only for installation purposes, but also for operation and maintenance purposes after the construction forces leave. A company should have its own standards to govern the creation and maintenance of electrical drawings. Such drawings are important to safety in that they help personnel understand the equipment on which they are working and how that equipment relates to the rest of the electrical system.

Perhaps the most important drawing from an electrical safety standpoint is the single-line or one-line drawing. The single-line drawing gives a quick overview of the power sources and the sequence of disconnecting means. It is also important for planning lockout/tagouts.

The electrical plan drawing is another document that is important for maintenance and future modifications after installation, since it shows the location of electrical equipment and disconnecting means. This drawing is particularly valuable to new personnel, contractors, and visitors who are unfamiliar with a facility. The plan drawing should also be the document that shows embedded electrical conduits and hidden wiring. Knowing the approximate location of such concealed hazards provides the opportunity to protect personnel against the unexpected.

It is important that such drawings, particularly the single-line drawings, be accurate and easily accessible at all times.

9.2.6 Equipment identification

Every facility should have a standard that covers the design and use of identification plates and tags for electrical equipment in a facility. Good identification is very important toward safely energizing and de-energizing equipment. In particular, proper identification is an extremely important element toward ensuring that equipment on which personnel work has been truly placed in an electrically safe work condition. An "electrically safe work condition" is defined in Chapter 10. More information about equipment identification can be found in Chapter 2.

The symbols used on drawings should be from a nationally recognized standard so that they are understood by both in-house and contractor personnel. IEEE has several such standards. (See 9.7.)

9.2.7 Grounding and bonding

It is extremely important to electrical safety that equipment and structures be properly grounded and bonded. All too often, these items are left as afterthoughts when new facilities and equipment are being installed. Grounding and bonding should be planned and designed with the same importance as the other parts of the electrical system.

The terms *grounded* and *bonded* are well defined in the NESC and in the NEC. Bonding is "the electrical interconnecting of conductive parts, designed to maintain a common electrical potential." This definition is self-explanatory, and implies that the conductive path should be adequately sized, and connections properly installed, in order to maintain a path with impedance as low as possible. The term *bonding* obviously is not exclusive to grounding systems.

Grounded means "connected to, or in contact with, the earth, or connected to some extended conductive body that serves in place of the earth." The earth or the other conductive body is known as the ground. When used as a verb, grounding is the act of establishing the aforementioned connection and conductive body. When used as an adjective, grounding describes something that is used to make the connection to ground.

There are two different kinds of permanent grounding relative to electrical work. The first kind is system grounding. System grounding is attaching at least one point of the normally current-carrying electrical path to ground, either solidly or through an impedance. This system ground affects performance of the electrical system, making it more stable and predictable. From a safety viewpoint, system grounding limits the potential difference between uninsulated objects in an area, helps limit the magnitude of overvoltages due to transients, and provides the reference point for the return of fault currents so that faults can be isolated quickly.

Equipment grounding is intended for safety purposes, and is the act of bonding all non-current-carrying conductive objects together to create a low-impedance conductive body (or path) to a ground reference. Among the safety purposes for equipment grounding are

a) Providing a path for the safe conduction of static and lightning currents, to reduce the possibility of fire or explosion;
b) Providing a permanent low-impedance path for fault currents to return to the system ground when these currents get off of the normal current-carrying path;
c) Providing a temporary low-impedance path as a protective measure in case of accidental energization during maintenance activities. This type of grounding is often called temporary personal protective grounding. Since this type of grounding is associated with safe work practices, it is discussed in Chapter 10.

More about grounding and bonding can be found in IEEE Std 80-1986, IEEE Std 142-1991, and IEEE Std 1100-1992.

9.3 Installation safety requirements

As was mentioned in 9.1, installation safety requirements are contained in several documents. The NEC is undoubtedly the document that is most familiar to electrical equipment installers. Installers should do the following things to ensure an electrically safe facility:

a) Install equipment in accordance with the operating diagrams, manufacturer's instructions, and other design documents.

b) Know the requirements of the NEC.

c) Install equipment in a neat and workmanlike manner.

d) Challenge design information that appears to disagree with the NEC.

9.4 Safety and fire protection inspections

After installation of new equipment and facilities or the major modification of existing equipment and facilities, a safety and fire protection inspection should be initiated by the facility custodian, whether it is required or not. It is also prudent, and often cost effective, to do periodic inspections while the installation or modification is being performed in order to catch problems before they become difficult to correct. At this time, one should look for exposed conductors, NRTL labels, NEC violations, easily understood identification, ease of operability and maintenance, tight working spaces, adequacy of lighting, and other things that might compromise safety. In some jurisdictions, it is also required by local law that official inspections be conducted by an authorized inspection agency.

Periodic safety inspections and audits should be conducted after operations begin to ensure that conditions remain as safe as they were initially.

9.5 Preplan for safe maintenance

The design of a facility and its electrical equipment should include consideration for future maintenance. In order to remain in good, safe condition, the electrical equipment and facilities must be maintained properly. Dust and dirt, damaged enclosures and components, corrosion, loose connections, and reduced operating clearances can be the cause of employee injuries. Some of these conditions can also lead to fire. A thorough, periodic preventive maintenance plan should be established as soon as new facilities and equipment are installed.

Local procedures should be created as soon as possible to cover the maintenance of electrical equipment. Most of this information can be obtained from recognized standards and manufacturers' literature. Proper operation and maintenance are important to electrical safety because when things do not function as designed or planned, the results may be unexpected. Many injuries and fatalities have occurred when the unexpected happened.

NFPA 70B-1998 is an excellent guide to recommended practices for maintenance of electrical equipment. It also contains the "why's" and the "wherefore's" of an electrical maintenance program, as well as guidance for maintaining and testing specific types of electrical equipment. In addition, it contains information in its appendix regarding the suggested frequencies for performance of maintenance and testing. This is a good document to review while facilities are being installed.

See more about the maintenance of electrical equipment in Chapters 5 and 6.

9.6 Repairs and replacement parts

When maintenance requires repairs or replacement parts, it is important to keep in mind the potential safety consequences of poor workmanship or "jury-rigged" fixes. These can negate safety features that were a part of the original design.

Qualified persons should perform repairs in accordance with the manufacturer's instructions and drawings. Using unqualified or unknown persons to perform repairs, without checking out their credentials, is asking for trouble. If there are no in-house qualified persons, there are many qualified service organizations, both independent types and those that are associated with a particular manufacturer. Before agreeing to any contracts, it is important to make sure that the contractor has good electrical safety qualifications. If the contractor's safety qualifications are unsatisfactory, the use of that contractor should be refused until they furnish an OSHA-compliant program. Accidents on the job site usually cause management to become involved, even though there were no company personnel hurt. This involvement may cause management a lot of aggravation.

Using the wrong replacement part can also negate the original safety features. Be sure to check manufacturer's literature for the proper replacement parts. Be aware that there are counterfeit parts on the market, and be careful not to use them. Know your suppliers and where they get their parts. When proper parts are no longer available, ask the original manufacturer for a recommendation. Of course, if the inquiry involves old equipment, a manufacturer will often suggest replacing it with their modern equipment. Many times, that would be a wise decision to avoid repetitive problems. If the budget is tight, however, the manufacturer will usually still help solve the problem another way. A qualified, reputable service and repair company can also be of help.

9.7 References

This chapter shall be used in conjunction with the following publications. If the following publications are superseded by an approved revision, the revision shall apply.

Accredited Standards Committee C2-1997, National Electrical Safety Code® (NESC®).[2]

CFR 29 Part 1910, Occupational Safety and Health Standards.[3]

CFR 29 Part 1926, Safety and Health Regulations for Construction.

IEEE Std 80-1986 (Reaff 1991), IEEE Guide for Safety in AC Substation Grounding.[4]

---

[2]The NESC is available from the Institute of Electrical and Electronics Engineers, 445 Hoes Lane, P.O. Box 1331, Piscataway, NJ 08855-1331, USA (http://www.standards.ieee.org/).

[3]CFR publications are available from the Superintendent of Documents, U.S. Government Printing Office, P.O. Box 37082, Washington, DC 20013-7082, USA.

[4]IEEE publications are available from the Institute of Electrical and Electronics Engineers, 445 Hoes Lane, P.O. Box 1331, Piscataway, NJ 08855-1331, USA (http://www.standards.ieee.org/).

IEEE Std 142-1991, IEEE Recommended Practice for Grounding of Industrial and Commercial Power Systems (*IEEE Green Book*).

IEEE 1100-1992, IEEE Recommended Practice for Powering and Grounding Sensitive Electronic Equipment (*IEEE Emerald Book*).

NFPA 70-1996, National Electrical Code® (NEC®).[5]

NFPA 70B-1998, Recommended Practice for Electrical Equipment Maintenance.[6]

NFPA 70E-1995, Standard for Electrical Safety Requirements for Employee Workplaces.

9.8 Bibliography

Additional information may be found in the following sources.

[B1] ANSI Y32.9-1972 (Reaff 1989), American National Standard Graphic Symbols for Electrical Wiring and Layout Diagrams Used in Architecture and Building Construction.[7]

[B2] IEEE Std 315-1975 (Reaff 1993), IEEE Graphic Symbols for Electrical and Electronics Diagrams (Including Reference Designation Letters).

[5]The NEC is available from Publications Sales, National Fire Protection Association, 1 Batterymarch Park, P.O. Box 9101, Quincy, MA 02269-9101, USA (http://www.nfpa.org/). Copies are also available from the Institute of Electrical and Electronics Engineers, 445 Hoes Lane, P.O. Box 1331, Piscataway, NJ 08855-1331, USA (http://www.standards.ieee.org/).

[6]NFPA publications are available from Publications Sales, National Fire Protection Association, 1 Batterymarch Park, P.O. Box 9101, Quincy, MA 02269-9101, USA (http://www.nfpa.org/).

[7]This standard is available from the Sales Department, American National Standards Institute, 11 West 42nd Street, 13th Floor, New York, NY 10036, USA (http://www.ansi.org/). Copies are also available from the IEEE as part of the Electrical and Electronics Graphic and Letter Symbols and Reference Designations Standards Collection (http://www.standards.ieee.org/).

Chapter 10
Safe electrical work practices

10.1 General discussion

Another major objective of the electrical safety program is to establish safe electrical work practices. No matter how well a facility is designed and built, careless actions of personnel can still result in injury or death. Safe practices involving electrical work are necessary in all workplaces in order to enable employees to recognize electrical hazards and avoid exposure to them. These days, excluding overhead lines, there are very few cases in which equipment is designed and installed without adequate physical protection against electrical hazards to personnel. Even the overhead lines are essentially "protected" from personnel due to their elevation. Most exposure comes during start-up and check-out, and during maintenance, troubleshooting, and modification.

Safe practices is the most important area of the electrical safety program on which to concentrate. A significantly greater number of injuries and fatalities are the result of poor or careless practices than the result of poor equipment conditions. Look back at some of the case histories in Chapter 7.

It is not the intention of this chapter to repeat all of the safe work practice information already found in other nationally recognized regulations, codes, and standards. This chapter will however, give an overview of the helpful information related to safe practices that can be found in those documents.

10.1.1 Acronyms and abbreviations

CPR cardiopulmonary resuscitation
NEC National Electrical Code®
NESC National Electrical Safety Code®
NFPA National Fire Protection Association
OSHA Occupational Safety and Health Administration

10.2 Training

The electrical safety program can document all of the greatest electrical safe practices in the world, but, if this information is not implemented by the persons who are exposed to the electrical hazards, the information is almost useless. This is where training comes in. Personnel should be trained to understand the content of the rules, why they exist, and how to implement them in the field. Federal law, in Occupation Safety and Health Administration (OSHA) regulation 29 CFR 1910.332,[1] requires that employees who face a risk of electrical shock that is not reduced to a safe level by electrical installation requirements be trained in, and be

[1]Information on references can be found in 10.6.

familiar with, electrical safety-related work practices that pertain to their respective job assignments. Training is permitted to be either in the classroom or on the job. Some of both would be best.

Training on safety-related work practices should cover all personnel, not just persons associated with the electrical end of the business. In this modern day of technology, so-called "unqualified" persons are surrounded by a lot of electrically powered utilization equipment. Even though they don't need to know the construction or how things operate internally, they do need to know the possible electrical hazards. Therefore, everyone in the workplace should have some degree of electrical safety training so that they can be qualified to perform their assigned tasks. Part of this qualification should cover the electrical safety aspects of their assignments.

People need to understand the reasons why they should follow electrical safety-related procedures, standards, and practices so that they will approach such work with the proper attitude.

As mentioned in 8.4.10, electrical personnel should also be trained in emergency procedures, such as methods of releasing victims, first aid, and cardiopulmonary resuscitation (CPR), since they might someday have a need for these techniques in their job assignments.

10.2.1 Qualified vs. unqualified persons

Among the first items discussed in NFPA 70E-1995, Part II, Chapter 2, are qualified and unqualified personnel. The definition of a qualified person, found in the introduction of NFPA 70E-1995, is "one familiar with the construction and operation of the equipment and the hazards involved." In order to be familiar with electrical equipment, a person should be trained, not only on the technical and mechanical aspects of the equipment or a specific work method, but also to avoid the electrical hazards of working on or near exposed energized parts. Obviously, an unqualified person is one who does not meet the criteria for whatever equipment is being worked on, or for the activity that is being performed.

The terms *qualified* and *unqualified* are often misunderstood. Some people think that to be electrical safety qualified, one must be a fully qualified electrician who has had a long list of electrical training courses. In reality, a person may be electrical safety qualified to perform only a limited number of tasks on or near specified electrical equipment. Take, for example, a tree trimmer. He or she certainly is not a fully qualified electrical worker as we might think of one, but he or she should be electrical safety qualified to work near overhead electric lines.

NFPA 70E-1995, Part II, Chapter 2, addresses the subject of training for qualified persons (i.e., electrical safety qualified). According to that document, they shall

- Be trained in, and knowledgeable of, the construction and operation of the equipment or a specific work method;
- Be trained to recognize and avoid the electrical hazards that might be present with respect to such equipment or work method;

— Be familiar with the proper use of special precautionary techniques, personal protective equipment, insulating and shielding materials, and insulated tools and test equipment.

Persons permitted to work on or near exposed conductors and circuit parts must also be trained in and familiar with

— The skills and techniques necessary to distinguish exposed live parts from other parts of electrical equipment;
— The skills and techniques necessary to determine the nominal voltage of exposed live parts;
— The approach distances specified in the table of NFPA 70E-1995, Part II, Chapter 2, and the corresponding voltages to which the qualified person will be exposed;

NOTE—The distances in Table S-5 of 29 CFR 1910.333(c) appear to pertain only to overhead lines. NFPA 70E-1995 attempts to also clarify safe approach distances for other kinds of electrical installations.)

— The decision-making process necessary to determine the degree and extent of the hazard and the personal protective equipment and job planning necessary to perform the task safely.

10.3 Electrical safety controls

Electrical safety controls include both administrative controls and self-controls. The administrative controls are specific facility or company rules regarding the conduct of work on or near electrical conductors and circuit parts. Administrative controls include many items, such as training and qualification of personnel, a work request system, job planning, work authorization documents, procedures, and audits. These subjects are discussed in several other parts of this guide. These controls are often thought of as paperwork, but they provide a logical, organized plan to address potentially hazardous electrical tasks. Self-controls are those steps that one takes to ensure one's own safety, and are discussed in 10.3.3.

10.3.1 Procedures

It is important to develop in-house standard procedures in order to provide direction to employees regarding the safety requirements and precautions to take while working on or near exposed conductors or circuit parts. This is especially true when personnel will be, or may be, exposed to energized electrical conductors and circuit parts. These procedures should be designed as easily used references for the kinds of tasks in which an employee could be exposed to an electrical hazard. These procedures should address work on electrical equipment at all voltage levels. In some facilities, these procedures are made requirements that carry penalties for violating them.

These procedures are the interface between the "planning" and the "doing." They are designed to provide an awareness of both electrical hazards and discipline for all personnel who are required to work in an energized electrical environment. A procedure on safe practices on or near electrical conductors allows for an instant audit of what is required to perform work on or near energized electrical conductors and circuit parts. A detailed procedure for the normal activity on each different type of major electrical equipment, and for the performance of electrical tests that may be done on a regular basis, helps identify hazards and enables them to be eliminated or controlled. Advance planning and preparation for the unexpected can minimize the detrimental effects of hazards that might still occur. Adherence to the procedures provides discipline for handling general work, and prepares personnel to handle the few unforeseen special jobs that may develop.

Safe work methods are stressed. No shortcuts or spur-of-the-moment activity is permitted. Work on or near energized conductors and circuit parts that develops, and which has not been previously identified by a procedure, should be reviewed, and a special procedure should be written prior to the performance of the work.

Both general and job-specific procedures are needed due to the large variety of potentially hazardous electrical tasks. A general procedure would cover such routine things as opening electrical equipment enclosure doors and taking voltage measurements. Job-specific procedures are needed for unusual or unexpected tasks, or for work on specific pieces of unusual electrical equipment.

All procedures should be prepared by qualified electrical personnel who are familiar with a given facility or plant following a standard format that addresses the controls of the electrical safety program. A typical outline should include information as shown in Exhibit 1.

Exhibit 1

Typical outline of an electrical safe practices procedure

- *Title.* The title identifies the specific equipment where the procedure applies.
- *Purpose.* The purpose is to identify the task to be performed.
- *Qualification.* The training and knowledge that qualified personnel shall possess in order to perform particular tasks are identified.
- *Hazard identification.* The hazards that were identified during development of the procedure are highlighted. These are the hazards that may not appear obvious to personnel performing work on or near the energized equipment.
- *Hazard classification.* The degree of risk, as defined by the hazard/risk analysis, is identified for the particular task to be performed.
- *Limits of approach.* The approach distances and restrictions are identified for personnel access around energized electrical equipment.
- *Safe work practices.* The controls that shall be in place prior to, and during the performance of, work on or near energized equipment are emphasized.
- *Personnel protective clothing and equipment.* The minimum types and amounts of protective clothing and equipment that are required by personnel to perform the tasks

described in the procedures are listed. Personnel performing the work shall wear the protective clothing at all times while performing the tasks identified in the procedure.

— *Test equipment and tools.* All the test equipment and tools that are required to perform the work described in this procedure are listed. The test equipment and tools shall be maintained and operated in accordance with the manufacturer's instructions.
— *Reference data.* The reference material used in the development of the procedure is listed. It includes the appropriate electrical single-line diagrams, equipment rating (voltage level), and manufacturer's operating instructions.
— *Procedure steps.* The steps required by qualified personnel wearing personal protective clothing and using the approved test equipment to perform specific tasks in a specified manner are identified.
— *Sketches.* Sketches are used, where necessary, to properly illustrate and elaborate specific tasks.

10.3.2 Work authorization

Before beginning any work, particularly in an existing operating facility, a person should receive a request to do the work from the custodian, thoroughly plan the job, review the job plan with the custodian, and obtain permission from the facility manager to proceed with the work. Some kind of work authorization document is advisable to ensure that everyone who may be affected is aware of what is going on. In addition to approvals, this document could contain a checklist of safety items that should be considered before proceeding with the work. The work authorization document forces people to think about the safety aspects of the job. This concept applies to all kinds of work, not just electrical.

Again, as was stated in Chapter 7, when performing electrical work in a country other than the U.S., make sure that any laws of the country that may be applicable to the job being performed are known.

10.3.3 Self-controls before each task

Electrical safety self-control is a process by which one performs his or her own safety analysis before beginning any task. This is the first step of a personal hazard/risk analysis. It can be accomplished by simply asking questions of oneself. If one can honestly answer "yes" to all of the following questions, he or she has done a good job of controlling his or her own safety. If one responds "no" to any of the questions, there is a safety concern that he or she should address before proceeding with the work.

a) Do I fully understand the scope of the task?
b) Am I trained and qualified to perform this task safely?
c) Have I performed this task before; if not, have I discussed the details with my supervisor?
d) Have I thought about possible hazards associated with this task and taken steps to protect myself against them?
e) Have I determined whether or not I will be near exposed energized parts?

f) If I am going to be exposed to energized parts, can they be put into an electrically safe work condition? [If "No," skip to item i).]

g) Did I verify, using appropriate protective and test equipment, that the conductors or equipment are in a de-energized state?

h) Have I applied a lockout/tagout device?

i) If I will be exposed to energized parts, do I know what voltage levels are involved?

j) Do I know the safe approach distance to protect against the electrical shock hazard?

k) Do I know the safe approach distance to protect against the electrical arc/flash hazard?

l) If a permit for energized work is required, have I obtained one?

m) Do I have the proper electrical personal protective equipment for this type of energized electrical work?

n) Do I have the appropriate voltage-rated tools and test equipment, in the proper working order, to perform this task?

o) Have I considered and controlled the following factors in my work environment?

   1) Close working quarters;
   2) High traffic areas;
   3) Intrusion/distraction by others;
   4) Flammable/explosive atmosphere;
   5) Wet location.

p) Do I understand that doing the job safely is more important than the time pressure to complete the job?

q) Do I feel that all of my safety concerns about performing this task have been answered?

This set of self-control questions makes the employee slow down and think about what he or she is going to do. Applying these controls can significantly reduce the probability of the employee being injured or killed while performing an electrical task.

10.3.4 Identifying hazardous tasks

All work on or near electrical equipment should be evaluated to determine whether or not the work exposes personnel to an electrical hazard, and, if so, what the magnitude of the hazard is. This evaluation is called a hazard/risk analysis. For example, a small task might be to put a nameplate on an electrical equipment enclosure door. Now, if the nameplate is put on using adhesive with the enclosure door closed, no hazard is apparent. However, if a person must drill into the enclosure or open the door to attach the nameplate with rivets or screws, there is exposure to energized conductors.

No matter how simple a task might first appear to be, a hazard/risk analysis and the electrical safety self-control questions should be applied. One example of a hazard/risk analysis is given in Appendix D of NFPA 70E-1995, Part II.

10.3.4.1 Typical hazardous tasks in electrical work

The following tasks are some examples of possible exposure to energized conductors:

a) Measuring, testing, and probing electrical system components;
b) Working near battery banks;
c) Opening electrical equipment enclosure doors or removing covers;
d) Inserting or pulling fuses;
e) Drilling, or otherwise penetrating, earth, walls, or floors;
f) Pulling conductors in raceways, cable trays, or enclosures;
g) Lifting leads or applying jumpers in control circuits;
h) Installing or removing temporary grounds;
i) Operating switches or circuit breakers;
j) Working inside electronic and communications equipment enclosures.

10.3.5 Evaluating the degree of hazard

Each of the tasks mentioned in 10.3.4.1 should be evaluated for the degree of electrical hazard involved. For example, opening an enclosure door of a 120 V control panel containing many relays and terminals does expose a person to an electrical hazard. But the probability of serious injury may be small due to the lower voltage and lower fault current capacity. Whereas, opening the door or cover of an energized, 13 800 V primary disconnect switch enclosure exposes a person to a much greater danger due to the higher voltage and larger fault current capacity. Different procedures, personal protective equipment, approvals, and attendance would be required.

10.3.6 Actions to eliminate, minimize, or control the hazard

Obviously, it would be most desirable to eliminate any hazard. The best way to do that is to rethink the purpose of the job and why it cannot be accomplished by establishing an electrically safe work condition (see 10.4.1). If the answer is "inconvenience" or "saving a little time," then the answer is not good enough. First of all, it would violate the intention of the OSHA laws (see 10.4.2). Secondly, the possible consequences if something went wrong should be considered (see 7.4).

When it is not possible or feasible to establish an electrically safe work condition, it is extremely important that work on or near exposed energized parts be thoroughly planned and strictly controlled. Work shall be done only by qualified personnel who have been trained to use safe practices and protective equipment.

10.3.7 Permit for energized work

In addition to work authorization documents, it is desirable to have a special permit system to give specific permission to work on or near energized electrical conductors or circuit parts.

This permit would list the following items that are required before working on or near exposed energized electrical conductors or circuit parts:

a) Justification for energized work;
b) Personal and other protective equipment;
c) Test equipment;
d) Tools;
e) Attendants;
f) Approvals;
g) Existing procedures to use;
h) Special notes.

In the past, this was called a *hot work* permit. Today, it is preferrably called a permit for energized work. (See 10.5.)

10.4 Working on or near de-energized equipment

The definition of the term *de-energized* can be found in IEEE Std 100-1996 and in several other documents. It is defined as "free from any electrical connection to a source of potential difference and from electric charge; not having a potential different from that of the earth."

At first thought, some people might think that they are safe if the electrical equipment on which they are going to work is de-energized. However, things are not always as they appear. The unexpected happens. A person should think, "What if...?." What if the wrong disconnect switch was opened? Or, since you can't watch the switch and work at the same time, what if someone turns the switch back on while you are busy working? What if a source of voltage from another circuit somehow gets accidentally connected onto the conductors on which you are going to work? What if a very large induced voltage is present? The point is that there are several things to consider to ensure a person's safety while working. De-energizing is only one part of creating an electrically safe work condition.

10.4.1 Establishing an electrically safe work condition

In the past, the methods that electrical personnel followed to protect themselves were lumped into a term called *clearance procedures.* In some cases, clearance simply meant permission to work on a particular system, whether it was energized or not. In other cases, clearance meant taking measures to ensure that equipment is de-energized, and to reinforce those measures with formal safeguards against altering that de-energized status for as long as clearance is required. The latter use of the word clearance is closer to the hazardous energy control requirements in place today. The term clearance is falling out of use in modern electrical safety terminology because it does not mean safety. Clearance (for work) is defined in 29 CFR 1910.269 as "authorization to perform specified work or permission to enter a restricted area." Today, for safety purposes, the phrase "establish an electrically safe work condition" is preferred. An electrically safe work condition is defined in Part II of NFPA 70E-1995. Section 2-3.1.3 of that document states

> "An electrically safe work condition shall be achieved and verified by the following process:
>
> a) Determine all possible sources of electrical supply to the specific equipment. Check applicable up-to-date drawings, diagrams, and identification tags.
>
> b) After properly interrupting the load current, open the disconnecting device(s) for each source.
>
> c) Where it is possible, visually verify that all blades of the disconnecting devices are fully open, or that drawout type circuit breakers are withdrawn to the fully disconnected position.
>
> d) Apply lockout/tagout devices in accordance with a documented and established policy.
>
> e) Use an adequately rated voltage detector to test each phase conductor or circuit part to verify that it is de-energized. Before and after each test, determine that the voltage detector is operating satisfactorily.
>
> f) Where the possibility of induced voltages or stored electrical energy exists, ground the phase conductors or circuit parts before touching them. Where it could be reasonably anticipated that the conductors or circuit parts being de-energized could contact other exposed energized conductors or circuit parts, apply ground connecting devices rated for the available fault duty."

When nondrawout, molded-case circuit breakers are being used as the disconnecting device mentioned in item b), visual verification of an open circuit, as suggested in item c), cannot be made. One technique that could be used to verify true opening is to have a voltmeter, or other voltage-indicating device, safely applied somewhere away from the breaker enclosure itself on the load side of the breaker before the breaker is opened. Always try to place the voltmeter at a point where exposure to energized conductors is minimized. Then, have someone watch the meter as the breaker is being opened. Simultaneous opening of the breaker and disappearance of voltage is generally a good indicator of disconnection. If that can't be done, the next best way is to measure load-side voltage (using safe practices and appropriate protective and test equipment), remove the meter, open the breaker, and measure again immediately. With multiple pole systems, all load-side poles should be verified to have voltage prior to disconnection. Again, apply a voltmeter to one of the poles. After the breaker is opened and the first pole is verified, move the meter, as safely and quickly as possible, to verify de-energization of the other poles.

CAUTION

1—Make sure to always test the functionality of the voltmeter after any verification for the absence of voltage.
2—Flash protection may be needed during these operations, depending upon the results of the hazard/risk analysis..

Working on or near electrical equipment without establishing the electrically safe work condition discussed in the preceding paragraph is taking the risk of having an electrical incident, injury, or fatality. Sometimes, such risks are necessary due to the nature of the task or the facility. Taking time to evaluate such risks and use the appropriate precautionary measures is also necessary.

A good basic rule to follow can be stated as follows:

All electrical circuit conductors, bare or insulated, are assumed to be energized until proven otherwise. They shall be de-energized, locked out, and tested for the absence of voltage before working on them or near them. Work on or near electrical circuit conductors and circuit parts may only be performed by trained and qualified personnel who have been authorized to do the work, using appropriate safe practices, personal protective equipment, tools, and test equipment.

Of course, from a practical standpoint, there is a need to have exceptions to this basic rule. For example, taking a voltage measurement is a common task during which a person is exposed to bare energized electrical conductors. Electrical safety measures for such exceptions should be thought about well in advance of any work, and made a part of the electrical safe practices procedures. Included in these measures should be a hazard/risk analysis, as explained in 10.3, and approval from appropriate levels of supervision and management.

10.4.2 Hazardous energy control (lockout/tagout program)

Hazardous energy control is not optional these days. It is required by law for all employees who work on de-energized equipment where there is potential for injury if the equipment is unexpectedly re-energized. This is an extremely important part of the overall electrical safety program, not only because it is the law, but also because it is a key effective method toward ensuring that employees have the electrically safe work condition described in 10.4.1. It is often called a lockout/tagout program.

OSHA regulation 29 CFR 1910.333(b)(2) states: "While any employee is exposed to contact with parts of fixed electrical equipment or circuits which have been deenergized, the circuits energizing the parts shall be locked out or tagged out or both in accordance with the requirements of this paragraph." That paragraph covers the following subjects:

a) Establishment and maintenance of written procedures for lockout/tagout;
b) Establishment of safe procedures for de-energizing equipment;
c) Requirements for the use of locks and tags;
d) Verification of the de-energized condition;
e) Requirements before re-energizing the circuits.

This means that a hazardous energy control program shall be established to cover all employees whose jobs could possibly expose them to energized electrical conductors or circuit parts.

Hazardous energy control of electrically operated equipment is important to nonelectrical workers also. Consider the following examples:

- Two mechanics working on a crane runway were knocked 40 ft to the floor below when a control-circuit failure caused the crane to start unexpectedly.
- A pipe fitter was scalded when an operator depressed the "open" button on a motor-operated valve.
- A two-man cleanup crew was buried in a storage silo when a conveyor was started accidentally.
- A hopper gate closed on the torso of a welder who was repairing the hopper lining.

All of these accidents have a common denominator. Although none involved electricians, nor electric shock or electrocution, all were electrically initiated. Furthermore, none would have occurred if proper electrical energy control procedures had been in effect.

A hazardous energy control procedure is a part of providing an electrically safe work condition for employees. This procedure is applicable to work on electrical equipment at all voltage levels, not just for higher voltage systems.

Hazardous energy control procedures, in the electrical business, are often referred to as lockout/tagout procedures. There are several existing documents in which lockout/tagout procedures are discussed in detail. ANSI Z244.1-1982 is a document that provides good guidance for establishing lockout/tagout procedures. ANSI Z244.1-1982 has a sample lockout/tagout procedure in its appendix.

It is quite obvious that the U.S. federal government is serious about control of hazardous energy in the workplace. OSHA regulation 29 CFR 1910.147 covers hazardous energy control in general and includes all kinds of hazardous energy, not just electrical. This document also contains a sample of a minimal lockout/tagout procedure. OSHA regulation 29 CFR 1910.333 is specifically aimed at lockout/tagout for electrical work in general industry. OSHA regulation 29 CFR 1926.417 discusses lockout and tagging of circuits for the construction industry. OSHA regulation 29 CFR 1910.269 discusses lockout/tagout requirements for power generation, transmission, and distribution type work.

Lockout/tagout practices and devices, including training, retraining, equipment, and procedures, are discussed in NFPA 70E-1995, Part II, Chapter 5.

Some detailed guidance toward establishing a lockout/tagout program is provided in 10.4.2.1 through 10.4.2.12.

10.4.2.1 A realistic approach to lockout/tagout

A lockout/tagout program should include provisions for issuing formal documentation (sometimes called a permit) to ensure that controls are in effect and cannot be removed until there is assurance that all personnel are no longer exposed to hazards. At first thought, it might appear that the ideal program would rigidly require that no work ever be undertaken on any equipment unless a permit is in effect. Such an inflexible policy would not only be unrealistic from an efficiency standpoint, but could create hazards equal to or greater than those that the program is intended to minimize. Immediate action may be necessary in an emergency. There may not be time to procure a formal permit. Therefore, the lockout/tagout program should include a provision that some small, noncomplex, or emergency jobs can utilize an individual nondocumented lockout/tagout. Conditions under which a nondocumented lockout/tagout can be used are given in 29 CFR 1910.147.

To be effective, a lockout/tagout program should be written in specific, rather than general, terms. Including unrealistic requirements or wording that is difficult to understand could damage the credibility of the program. A key element in the success of any lockout/tagout program is employee awareness that no violation of the program is acceptable.

A workable lockout/tagout program, then, is one that acknowledges reality without compromising safety. The program should not obstruct or delay work, but should provide an orderly method for expediting effective lockout/tagout.

10.4.2.2 Employee indoctrination and participation

All employees should be provided with, or have easy access to, a copy of the plant's lockout/tagout program. New employees should receive thorough indoctrination in the program. This indoctrination is, perhaps, even more important for production employees than for maintenance employees. Maintenance workers work intimately with the lockout/tagout program, acquiring familiarity in the course of their normal duties. For many production workers, though, direct involvement in the lockout/tagout program is minimal, and many accidents are the direct result of production employees operating equipment in violation of the plant's lockout/tagout program. For this reason, it is a good idea to devote regularly scheduled safety meetings as refreshers for the lockout/tagout program.

The plant's personnel policy should stipulate that any violation of the plant lockout/tagout program is considered a serious infraction of company rules and is subject to severe disciplinary action, up to and including termination of employment. More important, though, is obtaining the cooperation of employees to make the program work. Employees should be encouraged to feel that the lockout/tagout program is a personal tool for them to use to protect themselves from injury or death.

10.4.2.3 Padlocks and warning tags

Any lockout/tagout program should require that disconnect switches, circuit breakers, fuse holders, etc., be locked out and identified with a warning tag to indicate that the status of the equipment is not to be altered. Preferably, locking and tagging should be done by a qualified electrician, in conjunction with operations personnel, to ensure not only that the proper disconnecting means have been opened, but also that all the required operations have been performed. Such a policy ensures smooth, safe shutdown and restart.

Requiring that all locking and tagging operations be performed by qualified electricians, however, is unjustified in many types of manufacturing industries. This situation is especially true in light manufacturing industries because many jobs performed by nonelectrical personnel are of brief duration, and are on equipment that can be effectively made safe in a simple, straightforward fashion. With proper training, nonelectrical personnel can safely execute many simple lockout/tagouts.

Special instruction is essential for nonelectrical employees who are authorized to execute electrical lockout/tagouts. It is especially important to provide specific instructions in electrical safety procedures, emphasize practices to be avoided, and call attention to specific operations that do not effectively yield an electrically safe work condition.

Employees who are expected to execute lockout/tagouts should be issued a supply of padlocks and warning tags. If the lockout/tagout program allows or requires nonelectrical personnel to execute electrical lockout/tagouts, it should stipulate that assistance from the plant's electrical department be enlisted any time there is doubt as to what constitutes proper lockout/tagout.

10.4.2.4 Composition of the warning tag

Warning tags should be color-coded and prominently proclaim that the status of the equipment to which the tag is affixed must not be altered. ANSI Z535.5-1998 provides guidance on accident prevention tags. Space should be provided on the tag for the name of the person who applied the tag. Except as noted in 10.4.2.5, this person is the only one authorized to remove the tag. Space should also be provided on the tag for identifying the equipment that has been locked and tagged out, (e.g., No. 3 air compressor), and the time and date when the tag was placed. To minimize confusion in executing complex lockout/tagouts, an entry on the tag should identify the component of the electrical system to which the tag is affixed (e.g., 480 V air circuit breaker B-9). It is also desirable to provide spaces for indicating the nature of the work to be performed and for additional comments.

10.4.2.5 Personalized padlocks

All padlocks issued to an individual should be commonly keyed, but it is imperative that no two persons be issued padlock sets that can be operated by the same key. Designated supervisors can retain master keys for all padlocks. A documented procedure within the lockout/tagout program should clearly define the details for removing a lock and/or a tag when the person who installed them is not available. Personnel shall be trained on the use of such a procedure. The procedure shall include

a) An attempt to locate the person who installed the lockout/tagout device;
b) Verification that the person is not at the facility where the work is being done;
c) An attempt to contact the person, wherever he/she is, to inform him/her that the lock will be removed;
d) A method to guarantee that the person is informed, before returning to work at that facility, that their lock/tag has been removed.

10.4.2.6 Lockout/tagout permit

Some work requires rigid lockout/tagout control of the type that should not be the responsibility of the employee alone. Lockout/tagouts of this nature should be secured by a formal permit. This more formal approach is called a documented lockout/tagout. Typically, this type of lockout/tagout would be used on those types of jobs that are not simple and easily understood. Electrical work performed on medium- and high-voltage circuits is a good example. It would also include work on equipment that requires a complex lockout/tagout due to multiple sources of electrical energy. Also included would be jobs that require work inside of grinding mills, choppers, fan housings, ovens, storage tanks and silos, and similar situations in which personnel are in a position that unexpected equipment start-up would, without question, result in serious injury or death. In general, the documented lockout/tagout shall be used except when the conditions given in 29 CFR 1910.147 for a nondocumented lockout/tagout allow an exception.

No specific permit system can be recommended as good practice in all circumstances. A workable permit system can be developed only on an individual basis at the plant level by personnel intimately familiar with plant operations. Certain requirements that represent good practice in one plant might be inadequate or unworkable in another plant with different problems and a different personnel structure.

One fundamental feature, however, should be incorporated into any permit system. It should be designed with checks and balances. Specific responsibility for a particular operation should be assigned to an individual without relieving others of the obligation to double-check the status of the lockout/tagout before proceeding with their own assigned steps in the process. The permit system, then, should be developed to duplicate and reinforce, rather than dilute, responsibility.

Every step in processing a lockout/tagout permit, from the initial request to the official closing, should be confirmed in writing on an official form. The permit form should include spaces for every person involved to indicate the times and dates when the paperwork was received and when the action was taken. Completion of each step should be acknowledged by the signature of the person responsible for taking the appropriate action. Every person involved in processing the permit should be held responsible for checking the paperwork referred to them to see that everything is in order before proceeding with their own step.

10.4.2.7 Temporary release of lockout/tagout permit

In general, the lockout/tagout procedure should require that the equipment covered by the lockout/tagout permit not be altered unless the entire permit is closed out. Some types of work, however, require that equipment be operated to determine if the job is completed properly. For example, an employee balancing a large fan might be required to enter the fan housing repeatedly to attach balance weights, and the fan might be operated after each attempt at balance to obtain a vibration reading.

In such cases, it is permitted to remove the lockout/tagout devices temporarily and then replace them without going through all of the paperwork and approvals again. The need for the temporary release should be documented in the initial paperwork. In addition, everyone on the job shall be made aware at the beginning of the job, in the job briefing, that a temporary release will be employed. The physical actions of a temporary release are the same as if persons were removing their lockout/tagout devices permanently. Those persons who have to return to working on that job after the temporary release shall reapply their lockout/tagout devices before resuming any work.

10.4.2.8 Use of up-to-date single-line diagrams

The single-line diagram is the road map of an electrical system, tracing the flow of power from source to load. It indicates points at which power is fed into the system and at which power can be disconnected. No complex lockout/tagout should ever be attempted without first consulting the appropriate single-line diagram.

Because capacitors and instrument transformers are not normally viewed as power sources, they can be easily overlooked when securing a lockout/tagout. Yet, they can impose lethal voltages on the electrical system. Special care should be taken in checking the single-line diagram for the presence and locations of capacitors and instrument transformers any time a lockout/tagout is secured for performing work on electrical circuits.

In every facility, some person or group should be designated to keep single-line drawings up-to-date and audited. The location of drawings should be known and openly accessible to the personnel who are planning and performing work requiring a lockout/tagout. Drawings posted in substations and other locations throughout the plant should not be relied upon because they are often out-of-date.

10.4.2.9 Hazardous energy control for mechanical equipment

The lockout/tagout of valves, hydraulic and pneumatic operators, and engine and turbine prime movers is generally considered mechanical, rather than electrical. Such equipment normally is locked and tagged out by mechanical maintenance personnel or operating personnel. Valves, however, might be electrically operated, and pneumatic and hydraulic circuits are almost always electrically powered and are often electrically controlled.

Because the purpose of a lockout/tagout program is to provide for the total safety of all personnel, it is usually best to combine mechanical and electrical equipment on the same lockout/tagout permit. There may be cases, however, where it might be best to divorce the two. If different departments are responsible for mechanical and electrical lockout/tagouts, it is important to coordinate the two to ensure that electrically operated valves and hydraulic and pneumatic circuits are effectively locked out and tagged as a team effort.

10.4.2.10 Making the system workable

A workable lockout/tagout program should make some distinctions between the essential and the desirable. Balance should be struck between two extremes. At one extreme, there is no formal hazardous energy control system, and everyone must look out for themselves and make their own rules. At the other extreme, there is a rigid system that requires a formal lockout/tagout permit to be in effect before any work is undertaken on any equipment. The former is clearly unthinkable and in violation of federal law. The latter is not very cost effective.

Employee acceptance and respect are essential to the success of any lockout/tagout program. Imposing unrealistic requirements that are certain to be violated by employees, or ignored at the convenience of management, only fosters contempt for the entire system. A procedure should be developed that gives step-by-step instructions for implementing the program. The plant's current version of the lockout/tagout procedure should be considered to be inviolable. The program, however, should have a provision so that the procedure can be readily changed and revised when necessary. Copies of the latest revision to the lockout/tagout program should be issued to employees by identifying revision number and date. Training sessions on the changes should be held with all employees when changes are made. Superseded copies of the procedure should be collected or destroyed when the new version is issued.

The plant's lockout/tagout procedure should be developed with care, and a searching review of the procedure should be made periodically. Answers should be continually sought to the following questions:

- How might the procedure be improved?
- Are there loopholes that should be closed?
- Are all employees who are subject to exposure being adequately trained?
- Are there any requirements that are being tacitly ignored?
- If so, why are they being ignored?
- How can unrealistic requirements be modified to encourage compliance?
- Most importantly, does the procedure work?

10.4.2.11 Examples of poor hazardous energy control practices

The following items discuss some practices that were used in the past for safety control. These practices are *not* truly safe practices and should not be used today.

a) *Locking out a push-button, control switch, or other pilot device does not ensure that the circuit will remain de-energized.* A short circuit or ground in the control circuit can bypass the pilot device. Another employee might even engage the contactor or starter by hand. Unless the disconnecting means is opened and locked out, an employee should not place himself in a position where unexpected equipment start-up or energization might cause injury.

b) *Turning the handle of a disconnect switch to the "off" position does not ensure safety.* The switch linkage might be broken, leaving the switchblades engaged. Switchblades in the open position should be confirmed by visual inspection. The load side of the switch should also be checked with a voltage tester to ensure that the outgoing circuit is de-energized, and that there is no backfeed.

c) *Removing and tagging fuses does not constitute a lockout/tagout.* A lockout/tagout device should be attached to the fuse clips in a manner such that no fuses can be inserted without removing the device. If fuses are contained in a drawout fuse block, the tag should be attached to the fuse panel, not to the drawout block. Special precautions shall be taken to prevent shock whenever energized fuse clips that are accessible to the touch must be tagged.

d) *Simply opening a power circuit breaker does not ensure safety.* Even if the control fuses are removed, the breaker can still be engaged with the manual operating mechanism. The switchgear must be racked away from the bus contacts and into the "fully disconnected" position, and the racking mechanism shall be locked and tagged.

10.4.2.12 Other points to consider

The following is a list of points to consider when drafting an electrical lockout/tagout program:

a) The plant's lockout/tagout program should consist of two parts. The first part should cover general considerations. The second part should cover specific procedures that apply to documented or nondocumented lockout/tagouts.

b) Terms should be used in the same context in which they are used in the system. Definitions should be provided for lockout/tagout, lockout/tagout program, lockout/tagout procedure, lockout/tagout permit, released, restored, locked out, tagged, disconnecting means/disconnect switch, qualified, authorized, affected, and whatever other terms are appropriate in a particular facility.

c) No two crews working under different supervisors should be permitted to work under the same lockout/tagout permit. Each crew leader should secure a separate (redundant) permit.

d) Steps should be taken to eliminate the possibility that temporary grounding leads might be overlooked when energy is restored. Grounding leads should be issued only with a permit that identifies each set of leads by a distinctive number.

e) The lockout/tagout procedure should include detailed provisions for the lockout/tagout, grounding, and bleeding of capacitors and other stored energy devices.

f) The procedure should indicate that lockout/tagout alone does not necessarily indicate safety. All of the steps in establishing an electrically safe work condition should be applied.

g) Procedures for removing padlocks with a master key should be rigidly controlled. If the person who attached the padlock is absent from work, the steps in 10.4.2.5 shall be followed. If a time-card system exists in a facility, a suggested practice is to remove the employee's time card from the rack and substitute an official card informing them that their clearance has been abrogated and that they are to report immediately to a designated supervisor for briefing.

h) Provisions should be made for transferring lockout/tagouts from one person to another for work that must carry over from one shift to the next. The lockout/tagout program should provide for a written transfer procedure.

10.4.3 Temporary personal protective grounding

Sometimes, additional measures are desirable to provide an extra margin of safety assurance. Temporary personal protective grounds are used when working on de-energized electrical conductors to minimize the possibility of accidental re-energization from unexpected sources. Sometimes these are called *safety grounds* or *equipotential grounding.*

Induced voltages, capacitive recharging, and accidental contact with other circuits can occur. Depending on the electrical energy available, these occurrences could cause injury or death. More often, however, they only cause reflexive actions. For example, although most induced voltages will not normally cause serious injury themselves, they could cause a person to jump backward suddenly, possibly tripping against something or falling to the floor. Temporary protective grounding devices should be applied where such conditions might occur. Temporary personal protective grounds should be applied at possible points of re-energization. They can also be applied in such a way as to establish a zone of equipotential around a person. When these grounds are used, they shall be connected tightly, since they establish a deliberate fault point in the circuit. If current does somehow get onto the circuit, the grounds shall stay connected securely until a protective device clears the circuit.

It is difficult to set firm criteria for when temporary personal protective grounds are needed. Blanket requirements are usually established. Many times, it is a decision made in the field by the person performing the work. When there is uncertainty about exposure, it is wise to add this extra protection. Many industrial facilities and utilities require temporary personal protective grounding for all aerial power line work and for all work on power systems over 600 V because of the increased exposure these systems often have due to their length and location. Temporary personal protective grounding can also be used as the additional safety measure required when hazardous electrical energy control must be performed using a tag only.

Temporary personal protective grounding devices should meet the specifications in ASTM F855-96 and should be sized for the maximum available current of any possible event. Temporary personal protective grounds should only be installed after all other conditions of an electrically safe work condition have been established. Because the unexpected can happen at any time, however, the installation and removal of temporary grounding devices should be performed, by procedure, as the conductors are energized.

When installed inside equipment enclosures, temporary grounds should be lengthy enough to extend outside of the equipment so that they can be easily seen. If they cannot extend out, they should be made highly visible. Brightly colored tapes are helpful identifiers. Once they are installed, bare-hand work could be permitted.

It should be quite obvious that all personal protective grounds must be removed prior to re-energization. Identification and accountability controls may be necessary on large construction or maintenance jobs. The installation and removal of these grounding devices can be controlled by permit in order to avoid re-energizing equipment into a faulted condition.

The integrity of personal protective grounds should be maintained through the use of periodic inspection and testing. It is a good idea to document this inspection and testing.

10.5 Working on or near equipment that is, or can become, energized

A person's first reaction to working on or near electrical conductors or circuit parts should be to determine how to put them into an electrically safe work condition as described in 10.4.1. Occasionally, however, this is not practical. In some cases, de-energizing safety-related equipment might cause an even greater hazard than exposure to electricity. In other cases, de-energizing equipment in a continuous process might cause functional operating problems or involve great cost. In such cases, serious thought should be given to why the equipment can't be de-energized, and how the job can still be accomplished as safely as possible while energized. Essentially, the job can be done safely by personnel who have been trained, qualified, and authorized to use safe practices and appropriate personal protective equipment, tools, and test equipment. The key points listed below can be a safety checklist to determine whether or not one is ready to begin work.

a) Know the safe practices that are pertinent to the task that you will be performing.

b) A "permit for energized work" should be completed (see 10.3.7). This permit should include the justification and management approval for working on or near electrical conductors or circuit parts while they are energized.

c) Follow existing procedures where they exist. If previously prepared procedures do not exist, prepare a temporary job plan or procedure.

d) Stop work and rethink the situation if procedures can't be followed as written.

e) Be sure to perform an on-the-job hazard/risk analysis. (See 10.3.3 through 10.3.6.)

f) Know and maintain safe approach distances from exposed energized parts. (See 10.5.3.)

g) Obtain and use personal protective equipment for body parts that extend within the flash hazard distance and/or the shock hazard distance. (See Chapter 11.)

h) Make sure your standby person is present and knows never to leave you alone while you are exposed to the electrical hazard.

The terms *working on* and *working near* have always been subject to a wide variety of interpretations. In creating the 1995 revision to NFPA 70E, it was determined there was a strong need to define these terms. Working on was defined as "coming in contact with exposed energized electrical or circuit parts with hands, feet, or other body parts, with tools, probes, or with test equipment, regardless of the personal protective equipment a person is wearing."

Defining the term *working near* was not so easy. The word *near* called for establishing a boundary at some distance away from the exposed energized parts and giving the boundary and the space within it some names. The outermost boundary was called the limited approach boundary. Only electrical safety-qualified persons are permitted to cross that boundary and work in the contained space. Unqualified persons could only enter the space if accompanied by a qualified person, and then only for observation purposes. The term *working near* was then defined as "any activity inside the limited approach boundary of exposed energized electrical conductors or circuit parts that are not put into an electrically safe work condition." More will be discussed about approach boundaries and spaces in 10.5.3.

Another term that is difficult to define is *hot work*. It is old slang terminology for working on or near energized electrical conductors or circuit parts. There has been an attempt to discontinue the use of that term in the electrical business because it is too subject to interpretation, and it has so many meanings in other types of work. The term hot work could be interpreted to mean only working on energized conductors. Then again, it could be interpreted to mean working near, as well. Also, in general types of work, the term hot work means working on or near something that is thermally hot. In fire protection language, hot work is using devices and equipment that generate open flames, sparks, or excessive heat that could trigger a fire in an undesirable place. In radiological language, hot work means work with or near highly radioactive material. So, even though it's easier to say hot work, the term should be avoided. "Working on or near energized electrical equipment" is much more descriptive of the true hazard.

10.5.1 Overhead lines

Statistics on accidental electrocution show that quite a few of them involve work on or near overhead electric lines. Work on overhead lines is only to be done by qualified electrical lineworkers. Many times, due to the need to maintain service continuity, the lines are kept energized while work is being performed on them. Lineworkers must be well trained to perform such tasks using safe practices, appropriate personal protective equipment, and insulated tools.

When planning for work on overhead lines, however, one should always try to make the safest choice, which is to put the lines in an electrically safe work condition. Grounding the lines to create an equipotential zone within which a lineworker can be safe is advisable while working on overhead lines.

Work on or near overhead lines requires unique safety analysis because

a) The overhead lines can change position due to wind or other disturbances.

b) A person working on the lines is not usually in the most stable position.

c) The voltages and energy levels involved with overhead lines are often large.

Working near overhead lines, or near vehicles and equipment that could contact overhead lines, requires electrical safety training even for nonelectrical personnel. (See 10.5.2.)

The National Electrical Safety Code® (NESC®) (Accredited Standards Committee C2-1997) is a key document that gives significant detail regarding the safety rules for the installation and maintenance of overhead electric supply and communication lines. NFPA 70E-1995 also mentions safety around overhead lines in Part II.

The OSHA regulations that cover work on and near overhead electric lines are 29 CFR 1910.269 and 29 CFR 1910.333 for general industry, and 29 CFR 1926.955 for the construction industry.

10.5.2 Vehicles and mechanical equipment

NFPA 70E-1995 discusses requirements for the use of vehicles and mechanical equipment in the vicinity of overhead lines. To paraphrase, this document says that where it can be reasonably anticipated that parts of any vehicle or mechanical equipment structure will be elevated near energized overhead lines, they shall be operated so that the limited approach boundary distance (given in a table in Part II of NFPA 70E-1995) is maintained.

Electrical workers and others who are working near energized overhead electric lines should use insulated bucket trucks and other equipment that have insulated booms.

Many times, it is necessary to perform work near overhead lines, but not on them. This kind of work is done using mobile mechanical equipment that is capable of movement in an elevated position, such as cranes, derricks, aerial lifts, and dump trucks. In these cases, both the person operating the equipment and persons on the ground can be exposed to an electrical hazard. Persons on the ground should not touch the vehicle. Persons at ground level should not stand at or near the point at which the elevated equipment is connected to a grounding electrode.

OSHA regulation 29 CFR 1910.333 addresses vehicles and mechanical equipment in the vicinity of overhead lines. OSHA regulation 29 CFR 1926, Subpart V, also covers this situation for the construction industry.

The dimensions for clearances from overhead lines are given in several of the documents mentioned above. The National Fire Protection Association (NFPA) 70E Committee, however, is attempting to update requirements regarding safe approach distances, making them uniform and more easily understood. (See 10.5.3.)

10.5.3 Approach distances

In the past, approach distances to exposed energized electrical conductors were based on arc-over distance plus a generous safety factor. Today it is recognized that approach distances should take into account the flash hazard. Depending upon the amount of available fault energy, the safe limit of approach to protect against flash and blast could be further away than the distance necessary for shock protection.

Tables of approach distances from various sources in the past have given a variety of distances that were close to each other, but were not standardized. NFPA 70E-1995 has established a table that takes into account both shock and flash hazards, for both fixed equipment and equipment in which inadvertent movement is a factor. This table is located in Part II, Chapter 2, of that document. Also in the Appendix of Part II is a figure that shows the limits of approach and names the boundaries and spaces in the area that has been vaguely called "near."

NFPA 70E-1995 also provides a calculation that can be used under engineering supervision to determine the minimum approach distance to protect against flash instead of using the more conservative table values.

Other tables and discussions of approach distances can be found in the NESC and the OSHA regulations. At the time of this writing, however, those documents do not consider flash protection. As was mentioned in 7.2.2, several studies, tests, and technical papers are being written on the subject of the flash hazard. One of their objectives is to better define safe approach distances.

10.5.4 Switching operations

A major cause of personnel injury at industrial plants are the malfunctions that occur during the closing or opening of some types of switches or circuit breakers. Normally, switching operations are routine and everything goes smoothly. Every once in a while, however, something goes wrong. Some part of the electrical switchgear might come loose, a breaker compartment might be out of alignment due to wear, or interlocks might fail and load current might be initiated or interrupted. Also, due to increasing fault capacity in electrical power supply systems, some older, existing switches and breakers may not have sufficient capability to withstand possible fault currents. The failure of a fuse with insufficient fault-interrupting capability can likewise initiate a fault within a fusible switch enclosure. Such failure frequently initiates a phase-to-phase or ground fault, which either burns through the cover or blasts open the cover or door of the switch, thereby injuring or burning the person who operated the switch. Such an event is almost always unexpected. The person performing the switching, however, should always be prepared for the unexpected.

A well-defined procedure for closing switches and circuit breakers can go far toward eliminating the personnel hazards involved in this operation. For example, the method should state that no one should ever be allowed to stand directly in front of a switch or circuit breaker while it is being operated. The person performing the switching should stand off to the side of the switch or breaker enclosure, keeping the head and body as far as possible from the enclosure door. While performing the switching, the face should be turned away and the arm should be extended as much as possible to operate the switch. The use of a protective sleeve, or even a flash suit, should be considered, depending upon the required position of the person performing the switching, the available fault current, and whether or not trouble is suspected. Operation of the switch should be firm and quick; never an indecisive "teasing." Breakers and switches should be operated only with their doors and covers securely in place. Using this systematic method, if an electrical explosion does occur, a person's exposure to injury is reduced to a minimum.

Local procedures should be developed covering the safe operation of electrical equipment. Most of the information that is required to create a procedure can be obtained from the manufacturers' literature or by asking the manufacturer for a recommendation.

Proper operation is important for electrical safety because when things do not happen as expected, injuries or fatalities are more likely to occur. See more about the operation of electrical equipment in Chapters 2, 3, 4, and 12.

10.5.5 Penetrating into "unknown" space

A task that is encountered often, not only by electrical personnel, but by many employees in different crafts, is that of penetrating a wall, drilling into a floor, excavating the earth, or otherwise penetrating into a space containing unknown things. Many shocks, injuries, and quite a few deaths have resulted from performing such tasks without doing a thorough job of investigating what might possibly be in that space. These unknown areas might contain electric lines. When penetrating into unknown spaces, one should assume that there could be electric wires hidden in the space and, therefore, that extra precautions should be taken. A work authorization document should be the first thing required. (See 10.3.) This document should force some thinking before work actually starts. Other precautions that should be considered include checking the facility's structural, electrical, and underground line drawings; using detection equipment; using appropriate personal protective equipment; and using grounding devices on conductive tools and equipment. In some cases in which high-energy lines are suspected, even flash protection may be warranted. If the penetration is underground, one should be on the lookout for underground warning tapes or other indicators after the digging begins.

10.5.6 Other safe practices

There are many other safe practices mentioned in NFPA 70E-1995. Since most of them have become requirements of law through the OSHA regulations, NFPA 70E-1995 should be read and understood thoroughly. The following is an overview of some of those practices:

a) Employees should be alert while working on or near exposed energized parts. Supervisors should not allow them to work under such conditions if their alertness appears to be impaired, even temporarily.

b) Employees should not be permitted to work on or near exposed energized parts unless they have adequate illumination on the work area. They should also be instructed not to work around obstructions, or work in a manner in which their visibility is impaired. They should not reach blindly into an area that might contain exposed energized parts.

c) Employees working in a cramped or tight space that contains exposed energized parts are required to use protective shields, barriers, or insulating materials to avoid inadvertent contact with such parts. Doors and hinged panels shall be secured so that they cannot swing into an employee and bump him/her into the exposed parts.

d) Conductive materials and equipment, in an area where there might be exposed energized parts, shall be handled in such a manner that contact with the energized parts will be prevented.

e) Portable ladders used in areas with exposed energized parts shall have nonconductive siderails.

f) Safe practices also include using protective equipment, using tools properly, and employing other miscellaneous methods. These subjects are covered in Chapter 11.

g) Proper use of electrical equipment is also a safe practice and is discussed in Chapter 12.

10.6 References

This chapter shall be used in conjunction with the following publications. If the following publications are superseded by an approved revision, the revision shall apply.

Accredited Standards Committee C2-1997, National Electrical Safety Code® (NESC®).[2]

ANSI Z244.1-1982 (Reaff 1993), Safety Requirements for the Lock Out/Tag Out of Energy Sources.[3]

ANSI Z535.5-1998, Accident Prevention Tags.

ASTM F855-96, Standard Specifications for Temporary Protective Grounds to Be Used on De-energized Electric Power Lines and Equipment.[4]

CFR 29 Part 1910, Occupational Safety and Health Standards.[5]

CFR 29 Part 1926, Safety and Health Regulations for Construction.

IEEE Std 100-1996, IEEE Standard Dictionary of Electrical and Electronics Terms.[6]

NFPA 70-1996, National Electrical Code® (NEC®).[7]

NFPA 70E-1995, Standard for Electrical Safety Requirements for Employee Workplaces.[8]

10.7 Bibliography

Additional information may be found in the following source:

Palko, E., ed., “A realistic approach to electrical safety clearance procedures,” *Plant Engineering Magazine,* Feb. 17, 1977.

[2]The NESC is available from the Institute of Electrical and Electronics Engineers, 445 Hoes Lane, P.O. Box 1331, Piscataway, NJ 08855-1331, USA (http://www.standards.ieee.org/).

[3]ANSI publications are available from the Sales Department, American National Standards Institute, 11 West 42nd Street, 13th Floor, New York, NY 10036, USA (http://www.ansi.org/).

[4]ASTM publications are available from the American Society for Testing and Materials, 100 Barr Harbor Drive, West Conshohocken, PA 19428-2959, USA (http://www.astm.org/).

[5]CFR publications are available from the Superintendent of Documents, U.S. Government Printing Office, P.O. Box 37082, Washington, DC 20013-7082, USA.

[6]IEEE publications are available from the Institute of Electrical and Electronics Engineers, 445 Hoes Lane, P.O. Box 1331, Piscataway, NJ 08855-1331, USA (http://www.standards.ieee.org/).

[7]The NEC is available from Publications Sales, National Fire Protection Association, 1 Batterymarch Park, P.O. Box 9101, Quincy, MA 02269-9101, USA (http://www.nfpa.org/). Copies are also available from.the Institute of Electrical and Electronics Engineers, 445 Hoes Lane, P.O. Box 1331, Piscataway, NJ 08855-1331, USA (http://www.standards.ieee.org/).

[8]NFPA publications are available from Publications Sales, National Fire Protection Association, 1 Batterymarch Park, P.O. Box 9101, Quincy, MA 02269-9101, USA (http://www.nfpa.org/).

Chapter 11
Protective equipment, tools, and methods

11.1 Introduction

Electrical protective equipment serves to eliminate or reduce hazard severity, reduce the likelihood of an accident given that a hazard exists, and reduce the severity of the injury if an accident occurs. Historically, electrical protective clothing and conductor guarding were first applied to the prevention of electric shock injuries. In the 1970s, users and manufacturers began recognizing and addressing the electric arc hazard. In the early 1990s, Occupational Safety and Health Administration (OSHA) regulations and National Fire Protection Association (NFPA) standards began incorporating specific requirements to protect personnel from electric arc burns.

The selection of personnel protective equipment should be determined by a hazard analysis that determines the hazard severity and the parts of the body that could be exposed to the hazard. All body parts exposed to electrical hazards should be protected as a last line of defense from personal injury.

In addition to protective clothing, managing systems can be designed to augment the reduction of hazard exposure. Properly designed labeling and documentation practices serve to communicate and inform people of hazard presence and potential. Work practices can be engineered to reduce the potential for accident and injury.

The American National Standards Institute (ANSI) and the American Society for Testing and Materials (ASTM) standards that pertain to the selection, care, and use of protective clothing, equipment, and tools are summarized in Tables 3-3.6 and 3-4.11 of NFPA 70E-1995 [B9].[1]

11.1.1 Acronyms and abbreviations

ANSI American National Standards Institute

ASTM American Society for Testing and Materials

GFCI ground fault circuit interrupter

NFPA National Fire Protection Association

OSHA Occupational Health and Safety Administration

11.2 Personal protective equipment

Voltage-rated rubber gloves provide protection from hand contact with an energized source. Gloves are available in various voltage classes and with different cuff lengths. These gloves shall be used, inspected, and maintained to ensure their protective integrity. Leather protectors shall be used to prevent damage during use. Gloves shall be stored to prevent damage from sunlight, abuse, and contamination. They should be carefully visually inspected. The

[1]The numbers in brackets correspond to those of the bibliography in 11.8.

care, inspection, and testing of rubber gloves are detailed in ASTM D120-95 [B2]. Insulated rubber gloves are made in the following six different voltage classifications:

Class	AC proof test voltage (V)	AC maximum use voltage (V) (ASTM D120-95 [B2])
00	1 000	750
0	5 000	1 000
1	10 000	7 500
2	20 000	17 000
3	30 000	26 500
4	40 000	36 000

Most arc burns are incurred by personnel who are working close to energized parts and operating or servicing energized equipment that has available fault current sufficient to produce an explosive arc. Switchgear operation, hook-stick operation of fuses, or the repair or testing of components are typical activities that place people in such vulnerable positions. Suitable protection means for when the potential for severe faults is present include the following:

a) Leather gauntlet gloves;
b) Safety glasses;
c) Nonconductive hard hat;
d) Face-cover helmet;
e) Fire-retardant clothing or covering over normal clothes;
f) Greater separation of personnel by longer hook sticks, and shields;
g) Prohibiting work within the hazardous burn distances of energized parts by requiring the de-energizing of these parts before work is started.

The use of fire-retardant clothing, face shields and hoods, and leather gloves has improved the safety of the operators and has reduced the severity of injuries from explosive faults and electric arcs during the operation and servicing of electrical switchgear. The selection of arc-flash protective clothing is dependent on the severity of potential faults, the amount of incident energy that could be transferred to the person exposed, and the thermal characteristics of the protective clothing. NFPA 70E-1995 [B9] provides a method for determining fault severity.

11.3 Other protective equipment

11.3.1 Rubber blankets

Rubber blankets can be draped around conductors to provide a temporary insulating system. They were designed for utility open lines, but can be used in some industrial applications. Various clothespin-like clamps can be used to hold these devices in place. They have the same voltage classification as rubber gloves and should be inspected for holes before use. Outside influences should be considered when protecting the integrity of the electrical insulation.

11.3.2 Insulated tools and handling equipment

Insulated tools, ladders, and switchsticks provide protection from both shock and arc-flash burns. The tools and devices shall be rated for the voltage with which they may come in contact, and should be stored and inspected to maintain insulation integrity. Double-insulated electric power tools should be inspected and repaired according to the manufacturers' directions to ensure double-insulation integrity.

11.3.3 Doors, covers, shields, guards, and barriers

Doors, covers, shields, guards, and barriers serve to prevent contact with, or limit approach to, energized conductors. Their effectiveness in preventing unintentional contact with energized circuits is dependent upon the workers' understanding and conscious awareness of the guarded hazard. In situations in which equipment or circuits are isolated and locked-out in order to enable work on de-energized circuits, the doors, covers, shields, guards, and barriers that define the boundaries of the safe work area shall be understood. In addition to knowing what is locked-out, it is important to identify and communicate where energized sources may exist.

11.3.4 Ground fault circuit interrupters

Ground fault circuit interrupters (GFCI) sense when electric current flows through a person, and prevent serious injury by isolating power to the circuit within milliseconds. Although they were initially required only for use where portable tools, appliances, or equipment could be used in damp or wet locations, the expanded application of GFCIs offers a significant level of shock prevention to any use of temporary extension cords or portable tools and equipment.

11.4 Protective methods

Administrative controls or standard approaches to common tasks serve to minimize variables that contribute to operating errors. These may address common maintenance tasks, equipment design and installation, system documentation, and job planning.

11.4.1 Grounding of equipment

The integrity of equipment grounding is essential to personnel safety. Grounding is discussed in more detail in 9.2.7 and 10.4.3. Grounding is a primary protective method for protecting people from shock hazards that could exist in poorly grounded or ungrounded metallic raceways, equipment housing, or enclosures. The design and installation of equipment grounding is detailed in Article 250 of the National Electric Code® (NEC®) (NFPA 70-1996) [B8].

11.4.2 Alerting techniques

Examples of protective systems that serve to warn personnel of impending hazards include

a) Signs and placards;
b) Fences and other physical barriers;
c) Marking tape for underground lines;
d) Attendants.

11.4.3 Planning

Any work on or near energized electrical equipment has the potential for an accident that could result in serious injury or death, interruption of electric power, disruption of control systems, or damage to critical equipment. A risk analysis of the task to determine the accident potential and the consequences will help to make sure that the right decisions are made to ensure facility reliability and personnel safety. A good starting point for the prevention of accidents is to prohibit work on or near energized parts by requiring the de-energizing of these parts before work is started. Working on or near energized equipment should be managed as exceptions to the rule. NFPA 70E-1995, Part II, Chapter 2, and Appendices A, B, and D [B9], provide guidance for managing this approach.

Job plans or task procedures provide means to ensure that the hazards are recognized and managed prior to the beginning of work. If the task is relatively simple and involves one person, then it may not be necessary for the plan or procedure to be written. If, however, it involves more than one person, more than one craft, or extends beyond more than one shift, then it may be necessary to have a written plan in order to ensure the common understanding of all parties involved. In such instances, it is desirable to develop, in advance, a detailed plan for performing each step of the work safely. Such plans are most useful if they are written out in complete detail, with all involved crafts agreeing on each step of the procedure. Whether the plan is a mental plan, a verbal plan between two people, or a detailed written plan involving many people, effective planning and communication to all involved serves to achieve safe and error-free results. Guiding principles for effective planning are included in 8.4.3.2.

11.5 Drawings and other documentation

Drawings and other documentation are essential for identifying and communicating the information needed to plan and implement work on electrical systems with a high degree of safety for the people involved and reliability for the systems impacted. Additional information on drawings and documentation is included in 8.4.7, 9.2.5, and 10.4.2.8.

11.5.1 Safety electrical one-line diagrams

Safety one-line diagrams are made to show all the sources of electrical energy in an electrical power system. They are designed for the electrically safe operation of a power system only, and should have the following characteristics:

a) *Clarity.* The drawings should be easy to read in poor lighting conditions. Clarity should have priority over geographical location. If clarity requires that the systems be shown on two sheets, then that is the way it should be. The reason for this drawing is safety so the drawing shall be clear and legible.
b) *Correctness.* If the safety one-line drawing is not going to be kept correct, then it should not be made. There should be a system in the organization to keep the drawing correct, and to issue it to people who are authorized to do switching. These people shall be trained in the proper switching procedures.
c) *Component identification.* All power system components should be clearly identified on the drawing, as well as on the components themselves. Components are such things as switches, circuit breakers, cables, transformers, substations, potential transformers, etc. The components shall have only one identification, and it should be on both the drawing and the equipment itself. Engraved plastic signs or any means that do not fade with time are satisfactory. The identification signs should be large enough to be read at a distance and should not be painted over. They should be on all sides of the equipment being identified and placed where there can be no question that the component is indeed what the sign says it is. Any special warning such as "This is not a load break device" is acceptable, as long as the procedures are not on the identification sign. The purpose of this sign is to identify a component of a power system; it is not a substitute for trained people or a procedure manual. Short, alphanumeric designations are better than operating names. Avoid geographical descriptions (e.g., what do you call the "north acid pump" when another pump is placed north of it?). Avoid changing designations.
d) *Up-to-date, legible, and accessible.* The control of the diagram should include the destruction of the outdated drawings. A framed drawing on the substation wall is an excellent idea, as long as it is kept current and is replaced when it becomes faded or illegible. Drawings do fade with time, and the drawing should be legible under poor lighting conditions.

11.5.2 Panel schedules

Panel schedules fulfill the same purpose as safety electrical one-line diagrams. Maintaining the quality and accuracy of lighting and power distribution panel schedules should receive as much consideration as all other electrical distribution system diagrams.

11.5.3 Plot plans (location plan)

A plot-plan diagram is a necessary accompaniment to the one-line diagram for a complete description and mapping of the industrial and commercial electric distribution system. The system operators may be familiar with the location of the major components of the system, but total familiarity of circuit routing may not be available for some methods of circuit installation, particularly for installations that are out of view by normal observation methods.

Plot plans are important for a number of reasons, all of which could impact upon the operation of the industrial plant or commercial complex at some time. If a major catastrophe should occur, such as a fire or storm damage, a plot plan is a necessary tool if the distribution system is to be reconstructed. Expansion and rearrangement of an electrical distribution system could be extremely difficult without the knowledge of the location of existing system components. The plot plan can be important for identifying the proximity of electrical system components to other maintenance work that may be taking place.

11.6 Safety audits

An overview of why safety audits should be an integral part of the electrical safety program is found in 8.4.8. The general attributes of an effective audit are described below. One dictionary defines the term audit as "a formal examination and verification of financial accounts." The word, as used here, means the formal examination and verification of the safety program and practices for a specific power system or systems. The principal things that one should look for when conducting a safety audit are

a) *An operating one-line diagram.* A power system can be operated or maintained safely if there exist readily available drawings that show all the components of the power system. The drawing shall be correct, current, legible, and available to all those who operate or maintain the power system. There is no difference between an operating one-line diagram and the safety one-line diagram. Some operating one-line diagrams have information that is not required to operate the power system. There is no problem with adding this engineering-type information, as long as it does not distract from the clarity of the drawing. It is essential that the equipment be legibly marked with its operating name, and that this name be the same as the name on the safety one-line diagram. There shall be only one name for a power system component. More than one name for a piece of equipment can lead to switching errors and should not occur.
b) *Trained people operating and maintaining the power system.* Even a simple task, such as inserting a circuit breaker in its cubicle, is hazardous if the person performing it does not know the correct procedure. Inexperienced or untrained people are usually a major cause of most electrical accidents. This becomes more and more critical as the voltage increases. Medium-voltage (1001–69 000 V) equipment is much less forgiving than low-voltage (0–1000 V) equipment of switching or clearing errors.
c) *De-energized work procedures.* All conductors of electricity shall be considered energized until it has been proven that they are de-energized and grounded. This clearing procedure begins with the operation of all power source switches until there is no way that power can reach the part of the power system that people are working on without bridging a visible air gap. Specifically, do not work "behind" open circuit breakers. Voltage-sensing equipment that is tested before and after it senses the voltage should be used to determine if the part of the power system to be worked on is de-energized.
d) *Energized work procedures.* Written procedures should be developed for all energized work. These procedures should include a step-by-step outline of the work to be performed, protective equipment to be used, and familiarity with emergency-service procurement if a problem does occur. It should be approved by the person requesting that the energized work be done.
e) *Grounding.* Electrical power equipment should be grounded. Larger equipment, such as transformers and medium-voltage switchgear, have "pigtails" of 4/0 or larger wire to a grounding system of rods and interconnecting wire. This system needs to be

inspected for presence and condition. A periodic (not more than every five years) test program should be in place and documented.

f) *Corrosion.* No electrical system is safe if its components are corroded. Simply looking at equipment for corrosion can indicate whether it is safe for service.

g) *Maintenance practices.* There is a great deal of equipment in electrical systems that do not have to operate until there is a fault in the system. Protective relays, circuit breakers, control wiring, and current transformers are examples of such equipment. These devices simply do not do anything until a fault occurs, and then they have to work quickly and correctly. There are two ways to determine that they operate properly. The first way is to test them periodically. The second way is to throw a crescent wrench in the bus or wait until something else causes a fault. Doing this is not an acceptable method, is usually the most expensive way, and can cause fires, fatalities, etc. A relay/fuse short-circuit study should be accomplished for power systems when there are changes in the system, or at least every five years. Power company changes may affect the amount of short circuit that a plant has.

h) *Switching procedures.* All switching should be performed with written orders by people who are familiar with the equipment and the power system involved. Non-load-interrupting equipment should be specially marked, and its limitations should be adhered to in switching orders. See 10.5.4 and 12.4 for additional considerations.

11.6.1 Safety audit checklist

This checklist provides an assessment of the minimum requirements needed to safely operate and maintain electric power systems.

One-line diagram exists.	Yes ____	No ____
One-line diagram is legible.	Yes ____	No ____
One-line diagram is correct.	Yes ____	No ____
All persons who operate the power system have easy access to the current one-line diagram.	Yes ____	No ____
Equipment is labeled correctly, legibly, and in accordance with the one-line diagram.	Yes ____	No ____
Persons who operate/maintain electrical equipment are trained for the voltage-class equipment they operate/maintain.	Yes ____	No ____
De-energized procedures and equipment exist and are used.	Yes ____	No ____
Energized work procedures exist and are followed.	Yes ____	No ____
Equipment is grounded.	Yes ____	No ____
Ground system is tested periodically.	Yes ____	No ____
Electrical equipment is free from corrosion.	Yes ____	No ____
Proper maintenance practices are followed, especially for fault-protection equipment.	Yes ____	No ____
Recent (less than five years old) relay/fuse coordination study exists, and relays are calibrated to the setting recommended.	Yes ____	No ____
Power system is resistance grounded.	Yes ____	No ____
Written switching orders are used.	Yes ____	No ____

11.7 Safety morale

With all of the possible physical means available for working and operating safely, good results still will not be attained unless both workers and their supervisors believe in safety and work at it all the time. Some may assume the philosophy that "the other guys need safety, but not me; I'm good enough that I don't make mistakes." Such attitudes can be overcome by constant reminders that the safe way is the only way. Posters, meetings, and safety instructions, along with specific technical instructions on each job, are a few methods that should be used.

Working safely is a condition of employment and a basic responsibility of every employee. This responsibility is the basis for most of our safety standards and rules. "Do it safely, or don't do it."

All too frequently, it is the top management that needs to be convinced that safety must be a part of each job, even though it may not be particularly expedient. This is a difficult problem to overcome; however, if safety doesn't start at the top, it can never precipitate to the bottom. Just as product quality reflects management policy, accident rates reflect management's outlook toward personnel safety. Hence, with safety, it is obligatory to start at the top.

11.8 Bibliography

Additional information may be found in the following sources:

[B1] ANSI Z244.1-1982 (Reaff 1993), Safety Requirements for the Lock Out/Tag Out of Energy Sources.[2]

[B2] ASTM D120-95, Standard Specification for Rubber Insulating Gloves.[3]

[B3] CFR 29 Part 1910, Occupational Safety and Health Standards.[4]

[B4] *Guide to industrial electric power distribution.* Compiled by the editors of *Electrified Industry,* Chicago: B. J. Martin Co.

[B5] IEEE Std 141-1993, IEEE Recommended Practice for Electric Power Distribution for Industrial Plants *(IEEE Red Book).*[5]

[B6] Lee, R. H., "The other electrical hazard—Electric arc blast burns." IEEE TP IPSD 81-55.

---

[2]ANSI publications are available from the Sales Department, American National Standards Institute, 11 West 42nd Street, 13th Floor, New York, NY 10036, USA (http://www.ansi.org/).

[3]ASTM publications are available from the American Society for Testing and Materials, 100 Barr Harbor Drive, West Conshohocken, PA 19428-2959, USA (http://www.astm.org/).

[4]CFR publications are available from the Superintendent of Documents, U.S. Government Printing Office, P.O. Box 37082, Washington, DC 20013-7082, USA.

[5]IEEE publications are available from the Institute of Electrical and Electronics Engineers, 445 Hoes Lane, P.O. Box 1331, Piscataway, NJ 08855-1331, USA (http://www.standards.ieee.org/).

[B7] Lee, R. H., "The shattering effect of lightning—Pressure from heating of air by stroke current." *ICPS Conference*, Denver, CO, May 1985; ICPSD 85-32, May–June 1986.

[B8] NFPA 70-1996, National Electrical Code® (NEC®).[6]

[B9] NFPA 70E-1995, Standard for Electrical Safety Requirements for Employee Workplaces.[7]

[B10] "The Electric Power System," *Plant Engineering,* Oct. 15, 1981.

[6]The NEC is available from Publications Sales, National Fire Protection Association, 1 Batterymarch Park, P.O. Box 9101, Quincy, MA 02269-9101, USA (http://www.nfpa.org/). Copies are also available from.the Institute of Electrical and Electronics Engineers, 445 Hoes Lane, P.O. Box 1331, Piscataway, NJ 08855-1331, USA (http://www.standards.ieee.org/).

[7]NFPA publications are available from Publications Sales, National Fire Protection Association, 1 Batterymarch Park, P.O. Box 9101, Quincy, MA 02269-9101, USA (http://www.nfpa.org/).

Chapter 12
Safe use of electrical equipment

12.1 Introduction

In addition to the guidance provided in previous chapters, some helpful tips are provided here for the operation and use of common electrical equipment.

Portable electrical tools, temporary extension cords, and testing instruments are commonly used in any facility. The safe use of these devices and equipment is dependent upon the users' knowledge of both the task and specific equipment to be used, the integrity of facility grounding and protective systems [including the use of ground fault circuit interrupters (GFCIs)], and the systems in place to manage the inspection and maintenance of portable tools and equipment.

The operation of distribution, utilization, and control equipment is often performed by people without an in-depth knowledge of the electrical characteristics of the equipment. They should, however, be knowledgeable of the hazards involved in operating the equipment.

12.2 Portable electrical equipment

Portable tools, equipment, and appliances potentially expose everyone to electrical hazards in the workplace. Their safe use is dependent on both mechanical and electrical integrity, coupled with the users' awareness of potential defects. Considerations for the safe use of portable equipment include the following:

a) Cords and extension cords should provide an intact ground conductor from the building permanent wiring to the portable tool or equipment.
b) GFCIs provide a high degree of protection from electric shock, and should be applied where tools and portable equipment are used in potentially damp environments.
c) Portable cords, tools, and equipment should be maintained in a safe condition.
d) Portable cords, tools, and equipment should be inspected before each use.

Before taking any electrical tool or device into a classified area, the consequences of an unintentional arc or spark should be addressed. If the device is intrinsically safe, or is rated for use in the classified area, consideration should be given to ensure that the safety integrity is intact. If the device is not rated for use, then it shall be managed as a potential ignition source.

12.3 Test instruments and equipment

Test instruments are commonly used for verifying the absence or presence of voltage, for troubleshooting, and for obtaining diagnostic information. Their use involves working on or near energized circuits. The following are some considerations in managing the selection and use of test instruments and equipment:

a) Consider voltage test instruments as personal safety equipment rather than tools. These are the instruments that are used to determine whether or not a conductor is lethal or safe to touch. Treating them as safety equipment implies a higher level of attention and control.
b) Only use meters that are rated for industrial applications.
c) Minimize the number of manufacturers and models used in a facility.
d) As a minimum, instruments should meet the requirements of UL 1244-1993 [B3].[1]
e) Provide specific training for each model of the instrument. Each instrument has unique features, and even experienced personnel may be unfamiliar with subtle differences among similar models and styles.
f) Never use a damaged test lead or meter. Inspect leads for insulation damage or exposed metal before use. Never use a meter with obvious defects (cracked case, broken switch, water entry, etc.).
g) Reduce the risk of accidental contact by using leads with shrouded connectors and finger guards, and meters with recessed jacks. Consider the need for insulated gloves.
h) The high-voltage (typically 40 kV) accessory probes that are available from most manufacturers are for low-energy electronic circuits and not high-energy power circuits.
i) For personal safety, *always* test the meter on a known energized low-energy circuit (for example, 120 V general purpose outlet) both before and after making voltage checks. This verifies proper meter performance (battery, fuses, etc.).

12.4 Facilities infrastructure (power and light circuits)

No matter how experienced and knowledgeable personnel are, the person is the imperfect entity in the interaction between people and equipment. It is essential that deliberate attention be given to the planning and execution of switching operations, as it can compensate for deficiencies in knowledge or experience, distractions, and errors in judgment for any number of reasons. Switching activity that is performed in haste without a well-developed plan can be disastrous—both in terms of safety to personnel and the continuity of power to operations.

A well written procedure has the following features:

— Concisely and accurately describes the goal of the operation;
— Identifies unusual conditions;
— Provides a logical sequence;
— Accurately identifies equipment to be operated, including placement of tags and/or locks;
— Identifies vulnerable situations, including body position to minimize risk and personal protective equipment (PPE) to minimize injury if an accident occurs;
— Is reviewed by more than one knowledgeable person;
— Is reviewed, modified, and reviewed again if things do not go as planned.

[1]The numbers in brackets correspond to those of the bibliography in 12.5.

More than one person should be responsible for switching activity. Each step should be stated and then repeated for verification to ensure that the operation performed is correct and in sequence.

12.5 Bibliography

Additional information may be found in the following sources:

[B1] Nenninger, B. J., and Floyd, H. L., "Personnel safety and plant reliability considerations in the selection and use of voltage testing instruments," *IEEE Transactions on Industry Applications,* Jan./Feb. 1997.

[B2] NFPA 70-1996, National Electrical Code® (NEC®).[2]

[B3] UL 1244-1993, Electrical and Electronic Measuring and Testing Equipment (DoD).[3]

[2]The NEC is available from Publications Sales, National Fire Protection Association, 1 Batterymarch Park, P.O. Box 9101, Quincy, MA 02269-9101, USA (http://www.nfpa.org/). Copies are also available from.the Institute of Electrical and Electronics Engineers, 445 Hoes Lane, P.O. Box 1331, Piscataway, NJ 08855-1331, USA (http://www.standards.ieee.org/).

[3]UL standards are available from Global Engineering Documents, 15 Inverness Way East, Englewood, Colorado 80112, USA (http://global.ihs.com/).

INDEX

A

B

C

D

E

F

G

H

I

R

S

T

U

V

W